Graphing Calculator Manual

TI-83 Plus, TI-84 Plus, TI-89 and TI-Nspire

Patricia Humphrey
Georgia Southern University

The Triola Statistics Series:

Elementary Statistics, Eleventh Edition

Elementary Statistics Using Excel, Fourth Edition

Essentials of Statistics, Fourth Edition

Elementary Statistics Using the Graphing Calculator, Third Edition

Mario F. Triola
Dutchess Community College

Addison-Wesley
is an imprint of

Reproduced by Pearson Addison-Wesley from electronic files supplied by the author.

Publishing as Pearson Addison-Wesley, 75 Arlington Street, Boston, MA 02116.

Printed in the United States of America.

ISBN-13: 978-0-321-57061-1
ISBN-10: 0-321-57061-8

3 4 5 6 BB 11 10

Addison-Wesley
is an imprint of

www.pearsonhighered.com

Contents

1 Introduction to Statistics (and TI Graphing Calculators)

In this chapter we introduce our calculator companion to Triola's *Elementary Statistics* (11th ed.) by giving an overview of Texas Instruments' graphing calculators: the TI-83, -83+, -84+, -89, and Nspire. Read this chapter carefully in order to familiarize yourself with the keys and menus most utilized in this manual and the typical Introductory Statistics course.

You will also learn how to set the correct [MODE] on the calculators to ensure that you will obtain the same results as this companion does. You will learn other useful skills such as adjusting the screen contrast and checking the battery strength

Aside from the above technical skills, you will learn some basic skills that are particularly useful throughout your study of Triola's *Elementary Statistics*. Throughout this companion, we will present the uses of these calculators by illustrating their use on actual textbook examples. The first will be an exercise from the text which requires the selection of a random sample. There are not many calculator exercises in Chapter 1 of your textbook because it is an overview and introduction chapter. We will take the opportunity in Chapter 1 of this companion to introduce you to skills which you will find necessary throughout the other chapters. These skills include Home screen calculations and saving and editing lists of data in the STAT(istics) editor.

DIFFERENCES BETWEEN CALCULATOR MODELS

All TI graphing calculators have built-in statistical capabilities. Although a few statistical functions are "native" on the TI-89, most of the topics covered in a normal Statistics course require downloading the Statistics with List Editor application which is free at http://education.ti.com/us/product/tech/89/apps/appslist.html. Download requires the TI-Connect cable. This manual assumes the statistics application has been loaded on the calculator. If you have the newer TI-89 Titanium edition, the statistics application comes pre-loaded, and the TI-Connect cable is included with the calculator.

The TI-83 and -84 series calculators are essentially keystroke-for-keystroke compatible; however, the 84 does have some additional capabilities (some additional statistical distributions and tests, for example) with the latest version of the operating system, version 2.43 which is also available for download at education.ti.com. The regular TI-83 (not the plus version) does not have the ability to use APPS (applications) which are in some cases extensive programs. If you have one of these regular calculators, you will not be able to use the APPS included on the CD-Rom accompanying the text to load data sets, but will have to key them in yourself. (If you have the cable and TI-Connect software, these can be loaded from the .txt files included on the CD — see the Appendix for details). Regular TI-83 users will, however be able to use the programs on the CD for such applications as analysis of variance and multiple regression.

There are some key differences in calculator operation and menu systems which will, in some cases necessitate separate discussions of procedures for the different types of calculators. Some of these become apparent in the next section. Not only are there differences between the three series, but there is a key difference in operation between the TI-89 and the TI-89 Titanium edition. On the Titanium, all "functions" on the calculator are essentially applications—when a regular TI-89 is turned on, the user is on the "home screen" similar to that for the TI-83 and -84. When the Titanium edition is first turned on, one must scroll using the arrow keys to locate the desired application—we'll say more about this later.

The newest calculator is the TI-Nspire. This calculator works very differently with a document/page/problem "filing system." Although radically different in its menu structure, it is easy enough to follow along with the instructions for the other models after a bit of practice. This manual assumes you have the latest version of the operating system (1.4) installed. Download of these operating systems is also free at education.ti.com.

KEYBOARD AND NOTATION (TI-83/84/89)

All TI keyboards have 5 columns and 10 rows of primary keys (the Nspire adds additional keys devoted to alpha characters). This may seem like a lot, but the best way to familiarize yourself with the keyboard is to actually work with the calculator and learn out of necessity. The keyboard layout is identical on the 83+ and 84+, and differs from the 83 by the substitution of the [APPS] key for the [MATRX] key. The layout of the 89 (and 89 Titanium) keyboard is similar, but some functions have been relocated. You will find the following keys among the most useful and thus they are found in prominent positions on the keyboard.

- The cursor control keys [◄], [►], [▲] and [▼] are located toward the upper right of your keyboard (on the Nspire these are located on the "NavPad" in the upper center). These keys allow you to move the cursor on your screen in the direction which the arrow indicates.
- The [Y=] key is in the upper left of the 83 and 84 keyboards. It is utilized more in other types of mathematics courses (such as algebra) than in a statistics course; however you will use the *yellow* [2nd] function above it quite often. This is the STAT PLOT menu. We will discuss the yellow [2nd] functions shortly. On TI-89 calculators, the Y= application is accessed by pressing [♦][F1].
- The [ON] key is at the bottom left of the keyboard on all models. Its function is self-explanatory. To turn the calculator off, press [2nd][ON] ([ctrl][on] for the Nspire).
- The [ENTER] key is in the bottom right of the keyboard. You will usually need to press this key in order to have the calculator actually do what you have instructed it to do with your preceding keystrokes.
- The [GRAPH] key is in the upper right of the TI-83/84 keyboard. On the TI-89, GRAPH is [♦][F3].

As mentioned briefly above, most keys on the keyboard have more than one function. The primary function is marked on the key itself and the alternative functions are marked in color above the key. The actual color depends on the calculator model. Below you will be instructed on how to engage the functions which appear in color.

The [2nd] Key

The color of this key varies with calculator model. If you wish to engage a function that appears in the corresponding color above a key, you must first press the [2nd] key. You will know the second key is engaged when the cursor on your screen changes to a blinking ■. As an example, on a TI-83 or -84 if you wish to call the STAT PLOTS menu which is in color above the [Y=] key, you will press [2nd] [Y=].

The [ALPHA] Key

You will also see characters appearing in a second color above keys which are mostly letters of the alphabet. This is because there are some situations in which you will wish to name variables or lists and in doing so you will need to type the letters or names. If you wish to type a letter on the screen you must first press the [ALPHA] key. The color of this key also depends on the exact model model of calculator. On 83's and 84's it is directly under the [2nd] key; it is one place to the right of that on both 89's. You will know the [ALPHA] key has been engaged when the cursor on the screen turns into a blinking ▣. After pressing the [ALPHA] key you should press the key above which your letter appears. As an example if you wish to type the letter E on an 83 or 84, press [ALPHA] [SIN] (because E is above [SIN]. To get the same letter E on an 89, press [ALPHA][÷].

Note: If you have a sequence of letters to type, you will want to press [2nd] [ALPHA]. This will engage the colored function above the [ALPHA] key which is the A-LOCK function. It locks the calculator into the Alpha mode, so that you can repeatedly press keys and get the alpha character for each. Otherwise, you would have to press [ALPHA] before each letter. Press [ALPHA] again to release the calculator from the A-LOCK mode.

Some General Keyboard Patterns and Important Keys

1. The top row on 83's and 84's is for plotting and graphing. On 89's these functions are accessed by preceding the desired function with [♦].
2. The second row down has the important QUIT function ([2nd] [MODE] on 83's and 84's, [2nd][ESC] on 89's). On 83's and 84's it also contains the keys useful for editing ([DEL], [2nd] [DEL] (INS), [◄], [►], [▲] and [▼]). INS and DEL on 89's are both combination commands: INS is [2nd][←] and DEL is [♦][←].
3. The [MATH] key in the first column on 83's and 84's leads to a set of menus of mathematical functions. Several other mathematical functions (like [x^2]) have keys in the first column. On a TI-89, [2nd][5] leads to the Math menu.
4. The keys for arithmetic operations are in the last column ([÷] [×] [−] [+]).
   Note: On all input screens, the [÷] shows as /, and the [×] shows as *. On both 89 models, when the command is transferred to the display area the * is replaced with a · and division looks like a fraction.
5. The [STAT] key, on 83's and 84's will be basic to this course. Submenus from this key allow editing of lists, computation of statistics, and calculations for confidence intervals and statistical tests. On 89's with the statistics application, one starts the application using the key sequences [♦][APPS] and selecting the Statistics application. On the 89 Titanium, quit the current application ([2nd][ESC]) and locate the Stats/ListEd application. On 83's and 84's the second function of the [STAT] key is LIST. This key and its submenus allow one to access named lists and perform list operations and mathematics.
6. The [VARS] key on 83's and 84's allows one to access named variables. On TI-89's this is [2nd][−] which is named [VAR-LINK], and is used for both lists and variables.
7. [2nd][VARS] calls the distributions (Distr) menu. This is used for many probability calculations. To get this menu on a TI-89, press [F5] from within the Stats/ListEd application.
8. The [,] key is located in the sixth row directly above the [7] key on 83's and 84's, while on 89's it is above the [9] key. It is used quite often for grouping and separating parameters of commands.
9. The [STO►] key is used for storing values. It is located near the bottom left of the keyboard directly above the [ON] key on all the calculators. It appears as a → on the display screen.
10. The [(-)] key on the bottom row (to the left of [ENTER]) is the key used to denote negative numbers. It differs from the subtraction key [−].

Note: The [(-)] shows as ⁻ on the screen, smaller and higher than the subtraction sign.

KEYBOARD AND NOTATION (TI-Nspire)

The keyboard on this model has some striking differences from the other TI calculators. First is the addition of 42 more keys for alphabetic characters (no need for an [ALPHA] key), punctuation marks, and other symbols such as >, <, π, =, and *i*. Other keys function differently (or are relocated) from other models.

The (grab) Key

This key in the center of the "NavPad" is the "grabber." In most cases with menus it functions the same as (enter). It also allows you to grab a data point (or line) in a graph and dynamically move it. (**Important Note:** Moving a point also changes the data value in the data spreadsheet).

The (home) Key

Pressing this key allows you to add a new application on a new page in a document. It also allows you to view the catalog of saved documents (option 7), or create a new document (option 6).

The ctrl Key

This key functions much as the 2nd key on other calculators. It allows you to access items written in grey above the main function on many keys. In addition, there are several keystroke sequences using this key that are useful:

- ctrl on turns the calculator off.
- ctrl + darkens the screen contrast and ctrl − lightens it.
- ctrl 3 moves down a "page" in the data spreadsheet and ctrl 9 moves up a "page."
- ctrl ▸ and ctrl ◂ move forward (or back) a page in the document.
- ctrl ▴ changes the screen to view thumbnails of the pages in a document. This is the easiest way to delete a page. Use the arrow keys to highlight the page you want deleted, then press clear.

The menu Key

Pressing this key brings up menus of possible commands within each application. The options differ with the application.

The esc Key

Pressing this key allows you to "back out" of menus or command boxes.

SETTING THE CORRECT MODE

If your answers do not show as many decimal places as the ones shown in this companion or if you have difficulty matching any other output, check your MODE settings. Below we instruct on setting the best MODE settings for our work. These are the ones we have used throughout this companion.

On an 83 or 84, Press the MODE key (second row, second column). You should see a screen like one of those below. If your calculator has been used previously by you or someone else the highlighted choices may differ. If your screen does have different highlighted choices use the ▴ and ▾ keys to go to each row with a different choice and press ENTER when the blinking cursor is on the first choice in each row. This will highlight the first choice in each row. Continue until your screen looks exactly like the appropriate screen below. Press 2nd MODE (QUIT) to return to the Home Screen.

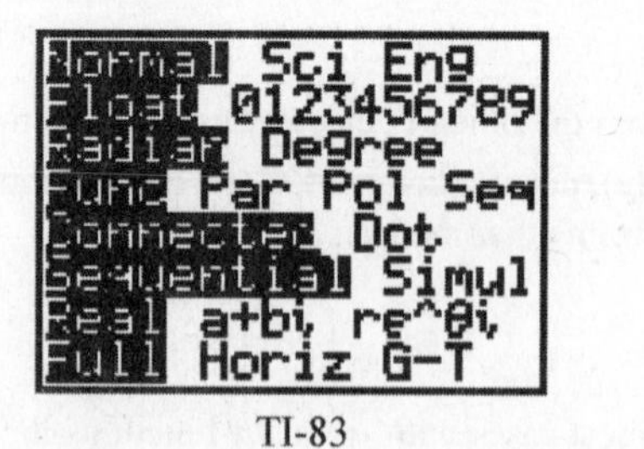

TI-83

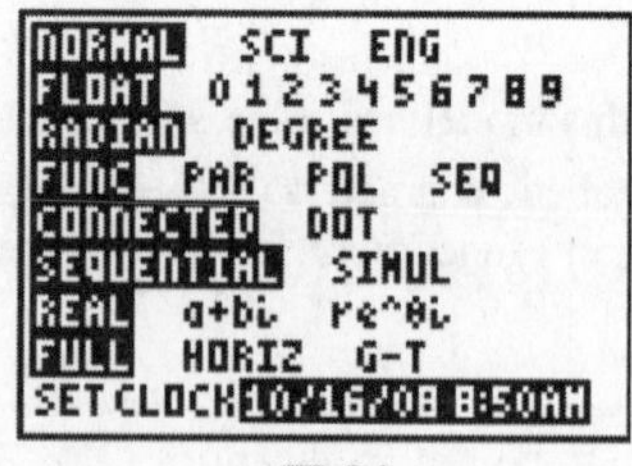

TI-84

On TI-89's, the default mode is to give "exact" answers. For statistical calculations, you will want to change the mode to give decimal approximations. To set this option, press MODE. Press F2 to proceed to the second page of settings, then arrow to Exact/Approx and use the right and down arrows to change the setting to 3:Approximate. Press ENTER to complete the set-up. The sequence of screens is shown below.

The TI-Nspire mode settings are accessed with option 8 on the home screen. You can set the mode for the calculator operating system, or just the current document. To move through the boxes, use the right (or left) arrows on the NavPad. Use the down arrow to expand a highlighted box, and make a selection; when finished, move so that OK is highlighted, then press [enter].

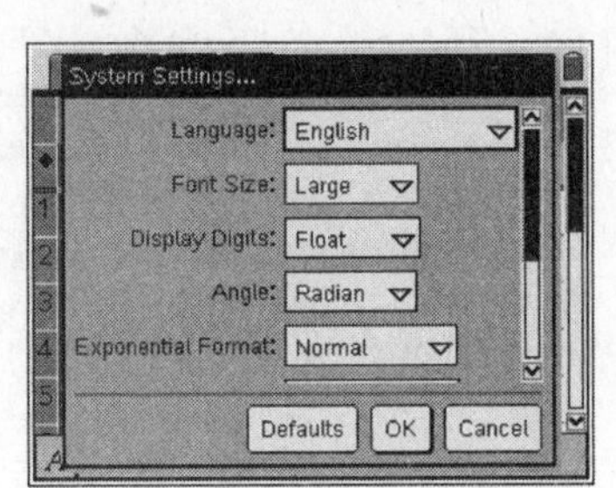

SCREEN CONTRAST ADJUSTMENT AND BATTERY CHECK

TI-83/84: To adjust the screen contrast, follow these steps:

To *increase* the contrast, press and release the [2nd] key and hold down the [▲] key. You will see the contrast increasing. There will be a number in the upper-left corner of the screen which increases from 0 (lightest) to 9 (darkest). To *decrease* the contrast, press and release the [2nd] key and hold down the [▼] key. You will see the contrast decreasing. The number in the upper-left corner of the screen will decrease as you hold. The lightest setting may appear as a blank screen. If this occurs, simply follow the instructions for increasing the contrast, and your display will reappear.

When the batteries are low, the display begins to dim (especially during calculations) and you must adjust to a higher contrast setting than you normally use. If you have to set the contrast setting to 9, you will soon need to replace the four AAA batteries. With newer versions of the operating system, your calculator will display a low-battery message to warn you when it is time. After you change batteries, you will need to readjust your contrast as explained above.

TI-89/Nspire:

To darken the screen (increase contrast), press [♦] (TI-89) or [ctrl] (Nspire) followed by [+]. To lighten the screen (decrease contrast) press [−] instead. The Nspire also has a battery status monitor – to check battery status, from the [home] screen, press [8] for `System Info`, then [3] for `Handheld Status`.

Note: It is important to turn off your calculator and change the batteries as soon as you see the "low battery" message in order to avoid loss of your data or corruption of calculator memory. Change batteries as quickly as possible. Failure to do so may result in the calculator resetting memory to factory defaults (losing any data or options which have been set).

A SPECIAL WORD ABOUT THE TI-89 TITANIUM

On the TI-89 Titanium, most all important functions which on other calculators are accessed by keystrokes, are applications (`Apps`). When the calculator is first turned on, you will be presented with a graphical menu of these applications, as at right. Paging through the screen to find the one you want can be tiresome and time consuming. There is a way to customize this screen so that you only see those applications you want to see.

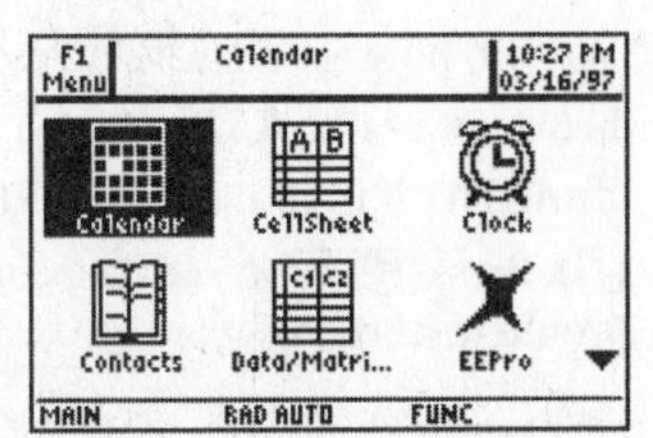

On the screen above, press [F1]. Press the right arrow key to expand menu selection `1:Edit Categories`. You will be presented with a list of possible catefories. Press [3] to select option `3:Math`.

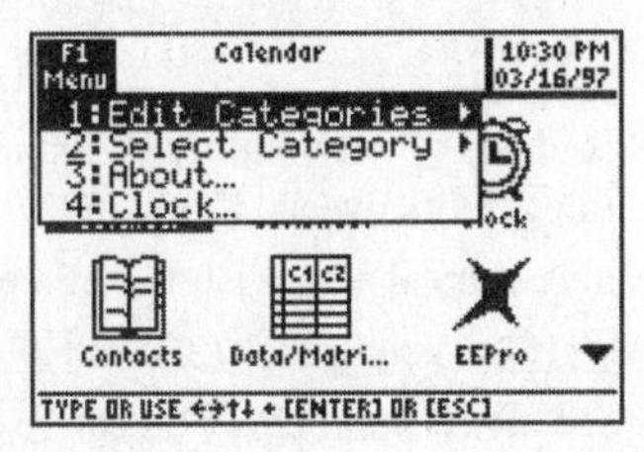

On this screen, use the down arrow to page through the list of apps. When you find one you want to be displayed, press the right arrow key to place a checkmark in the box. The screen at right shows that the Data/Matrix Editor and the Home screen have been selected. For ths statistics course, you will want these apps, along with the Stats/List Editor and Y= apps. Press [ENTER] when you have finished making your selections.

On this calculator, pressing [2nd][ESC] (Quit) will return you to the apps selection screen. There are two useful shortcuts between apps. The first is pressing the [HOME] key, which takes you directly to the Home screen. The other useful shortcut is pressing [2nd][APPS] which allows you to toggle between two apps.

RANDOM SAMPLES

All TI calculators have random number generators built into them. Such programs are dependent upon a value known as the "seed". Every random number you generate resets the seed, but we can set a particular value so that the sequence of random numbers will be the same. In practice, when one wants truly "random" numbers, one would eliminate this step. In this companion, we reset the seed each time we are generating a random sample, so that your output will match that in the companion. In normal practice, you do not need to reset the seed.

Random Sample and Simple Random Sample.

Picture a classroom with 60 students arranged in six rows of 10 students each. Use the calculator to simulate the results when a sample of 10 students is chosen as follows:
a) The professor will roll a six-sided die in order to choose a row. The students in that row are the sample.
b) The professor will choose a *simple random sample* of 10 students from the class.

TI-83/84 Procedure

On the Home screen, press the [CLEAR] key if the cursor is not in the upper left corner.

We will answer first part of the question by simulating the roll of a six-sided die. This means we need to generate a random integer between 1 and 6. Let's set the seed this time as follows:

a) On the Home screen type 123. Then press [STO▸] [MATH] [◂]. Watch what is happening – You went to the Math, PRB menu. Look at option 1. It is rand. You should see the screen at right.

b) Now press [1] and you will see that rand is pasted at the top of the Home screen as in the screen below.

c) Press [ENTER] and you will see the 123 as on the second line. This indicates the seed is now set.

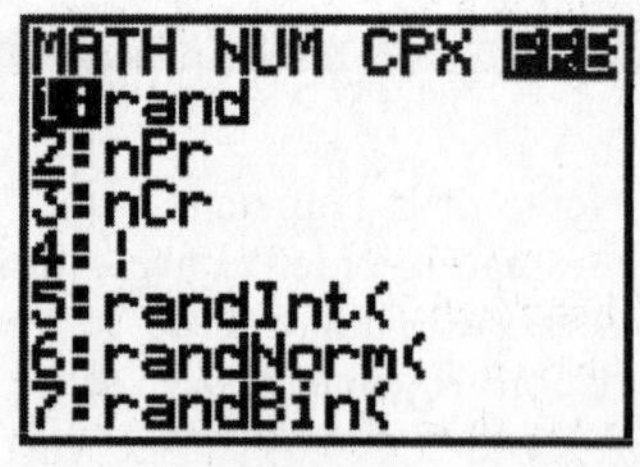

We will now simulate a single roll of a six-sided die by asking for a random integer between 1 and 6. Do so by pressing [MATH] [◂] [5] (thus choosing the option randInt(from the Math, PRB menu). Type 1,6 (remember the [,] is above the [7]). Press [ENTER] to see the result. We see the result is a 5, so the students in row 5 would be chosen.

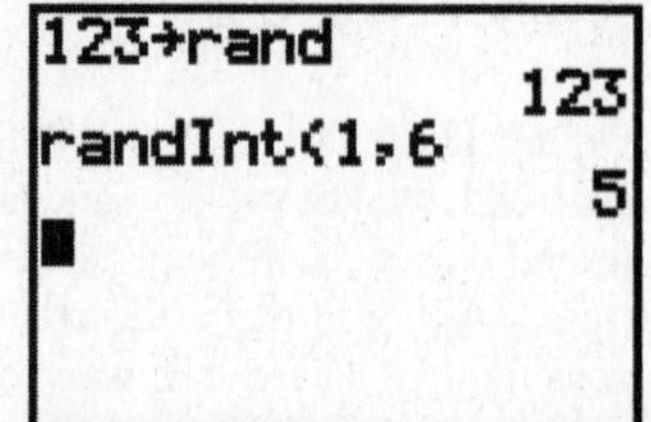

Now for part b of the question. We want a simple random sample of 10 of the 60 students. First number the students (1 to 60). We want a random sample of 10 integers between 1 and 60. This time we set the seed as 222. (This is only necessary if you wish to replicate our results.) After resetting the seed, press [MATH] [◂] [5]. Type 1,60,10. Press [ENTER]. You should see the screen at right. We cannot see all 10 numbers. You can use the [▸] key to move across and see the ones which

222→rand
222
randInt(1,60,10
... 1 2 11 27 52 ...

are not displayed. Note that the 10 we typed indicated we wanted a sample of size 10. In the above example, students 31, 1, 2, 11, 27, 52, 47, 4, 15, and 46 were chosen. It is possible that a student's number could have come up twice in the sample. If this occurs in a setting where this is not allowed (as in this one — a student should only be selected once), simply generate some more numbers until you have the desired sample size.

TI-89 Procedure

Load or select the `Stats/ListEd` application. Place the cursor at the top of an empty list. Press F4 (`Calc`).

We will answer part a of the question by simulating the roll of a six-sided die. This means we need to generate a random integer between 1 and 6. Let us set the seed as follows:

Arrow to `4:Probability` and press the right arrow to display the submenu. The menu option to set the seed is `A:RandSeed`. The easiest way to get there is to press the up arrow. Once the cursor highlights this option, press ENTER to select it.

Type the desired seed number. Here, I have used 123. Press ENTER twice to store the seed and return to the list editor.

Press F4 again, and display the `Probability` menu. This time we want to select menu option `5:randInt(`. Now type 1,6) (remember the , is above the 9 and we must close the parentheses). Press ENTER to see the result. We see the result is a 5, so the students in row 5 would be chosen.

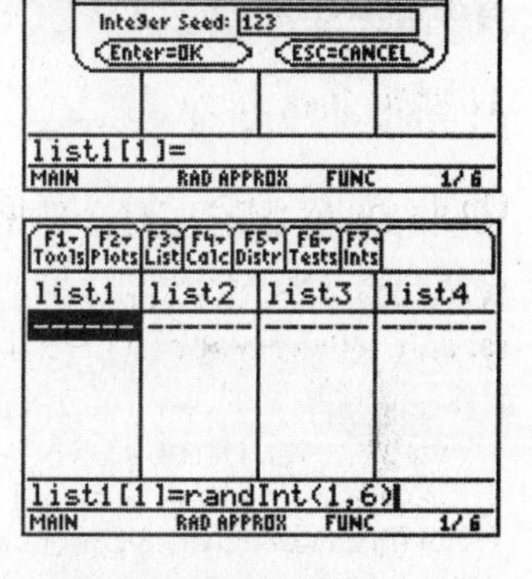

Next we tackle part b. We want a simple random sample of 10 of the 60 students. First number the students (1 to 60). Now you need a random sample of 10 integers between 1 and 60. As above, we set the seed to be 222. (This is only necessary if you wish to replicate our results.) With the cursor highlighting the *name* of the list, we now again find the `RandInt` command and change the parameters to 1,60,10) and press ENTER. This indicated we want 10 random numbers between 1 and 60.

Pressing ENTER populates the list. In the above example, students 31, 1, 2, 11, 27, 52, 47, 4, 15, and 46 were chosen. To see the entire list, scroll through it to the end. It is possible that a student's number could have come up twice in the sample. If this occurs in a setting where this is not allowed (as in this one), simply generate some more numbers until you have the desired sample size.

TI-Nspire Procedure

Random integer commands are located in the Probability section of the catalog. For a single random integer, locate the cursor in an empty cell of the spreadsheet. Type an equals sign (it's located at the upper right on the keyboard), then press the catalog key near the upper right of the keyboard. Move the cursor to highlight Probability and press the right arrow key to expand this section. Arrow down to `Random` and press the right arrow key again to expand that section. Now arrow down to highlight Integer and press enter. Enter the low and high numbers (1 and 6 in our example), then press [enter].

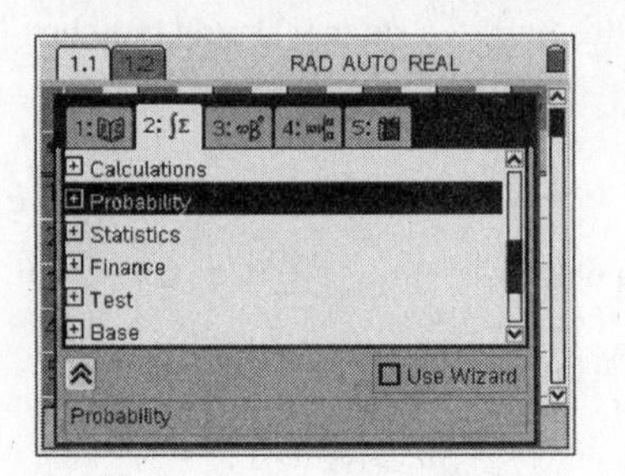

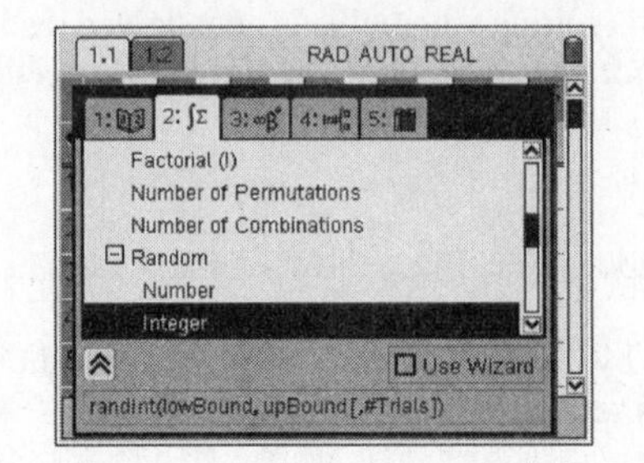

	A	B	C	D
•		=rand		
1	1987	2	2	
2	12	1		
3	233	1		
4	56	1		
5	1	4		

C1 =randInt(1,6)

You can generate a selection of more than one random number entering the command in the formula row. At right is our command to randomly select 10 of our 60 students.

	A	B	C	D
•		=rand		•0,10)
1	1987	2	2	
2	12	1		
3	233	1		
4	56	1		
5	1	4		

D =randInt(1,60,10)

HOME SCREEN CALCULATIONS

Cumulative Review Exercises: Calculator Warm-ups

We will use exercises 3 and 8 of the cumulative review at the end of Chapter 1 to illustrate some techniques. These examples also point out the importance of correctly using parentheses in calculations.

Exercise 3: $\dfrac{85-80}{3.3}$ (Standardize Shaquille O'Neal's height.)

We will calculate the value in two ways. In doing so, we will intentionally make a mistake to show you how to correct errors using the [DEL] key. We will also discuss the Ans and Last Entry features.

Type 86-8.0 (two intentional mistakes)

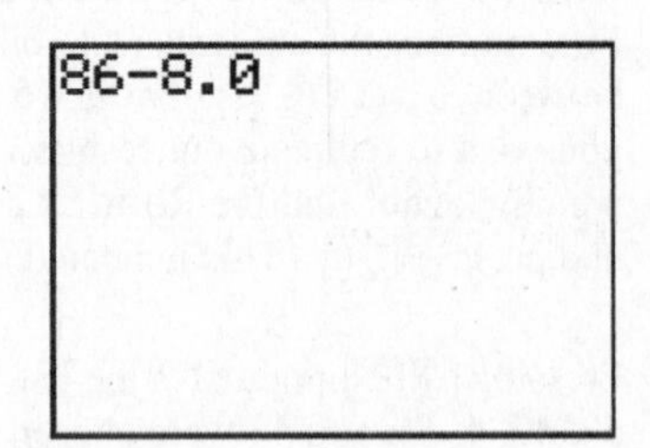

To correct the first mistake use the [◄] cursor key to move backward until your cursor is blinking on the decimal point. Press [DEL] (on an 89, [♦][←] or position the cursor to the right of the character to be deleted and press [←]) and the duplicate decimal point will be deleted. Now press [◄] until the cursor is blinking on the 6. Type a 5, and it will replace the incorrect 6. On an 89, move the cursor to the right of the error, press [←] and then type the correct 5. Press [ENTER] for the numerator difference of 5 as shown in the top of the screen below.

```
85-80
                5
Ans/3.3
      1.515151515
```

Press [÷]. (Note that "Ans/" appears on the screen). Type 3.3 and press [ENTER] for the result of 1.515.

Note: Ans represents the last result of a calculation that was displayed alone and right-justified on the Home screen. Pressing [÷] without first typing a value called for something to be divided, so Ans was supplied.

Press [2nd] [ENTER]. This calls the "last entry" to the screen. (in this case Ans/3.3). Press [2nd] [ENTER] again to get back to 85-80. Press the [▲] key to move to the front of the line. On an 89, press [2nd][◄].

On an 83 or 84, you will need to press [2nd] [DEL] (for INS); 89's are always in insert mode. You will see a blinking underline cursor. Type [(] to insert a left parenthesis before the first 9. Press [▼] ([2nd][►] on an 89) to jump to the end of the line. Type [)] [÷] 3.3 to see the result. Press [ENTER] for the same result as before.

Exercise 8: Compute the standard deviation of three body temperatures as

$$\sqrt{\frac{(98.0-98.4)^2+(98.6-98.4)^2+(98.6-98.4)^2}{3-1}}.$$

In this example, we will use the ANS function, illustrate syntax errors and show how to store quantities using variable names.

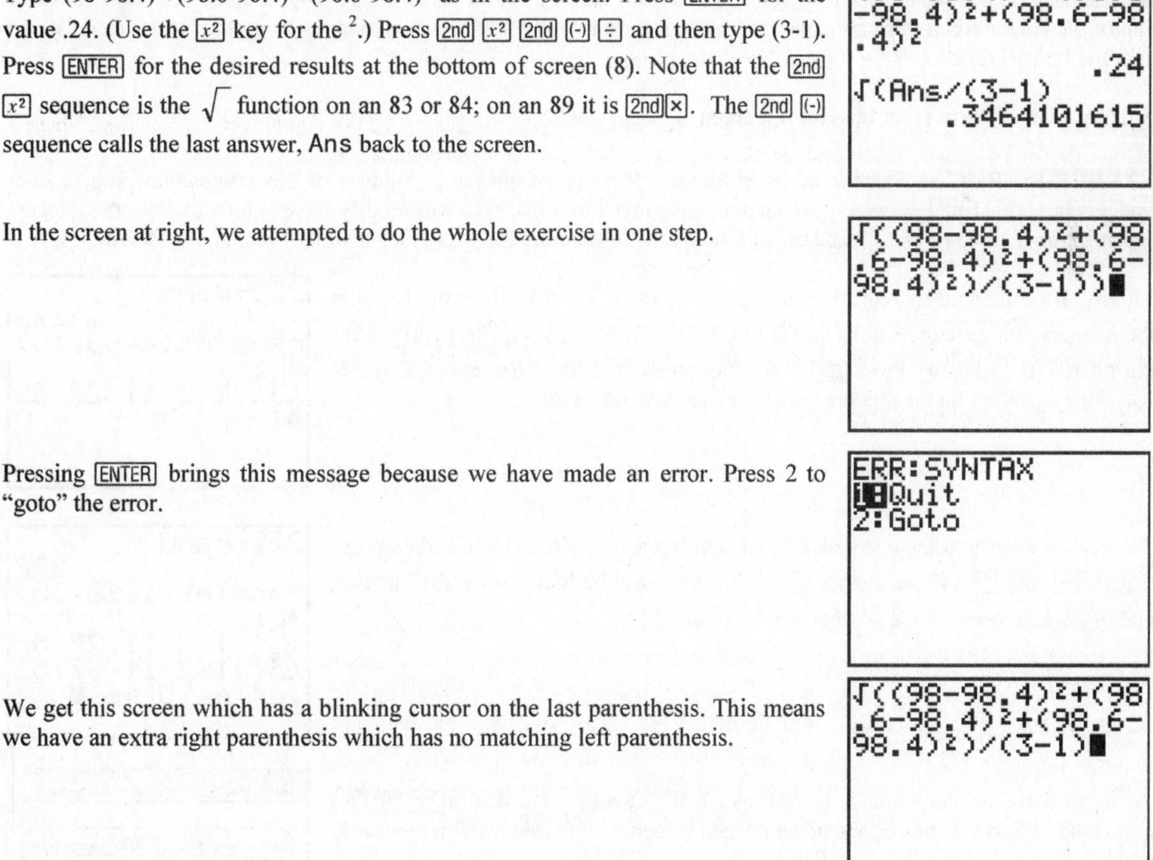

Type $(98\text{-}98.4)^2+(98.6\text{-}98.4)^2+(98.6\text{-}98.4)^2$ as in the screen. Press [ENTER] for the value .24. (Use the [x^2] key for the ^{2}.) Press [2nd] [x^2] [2nd] [(-)] [÷] and then type (3-1). Press [ENTER] for the desired results at the bottom of screen (8). Note that the [2nd] [x^2] sequence is the $\sqrt{\ }$ function on an 83 or 84; on an 89 it is [2nd][×]. The [2nd] [(-)] sequence calls the last answer, Ans back to the screen.

In the screen at right, we attempted to do the whole exercise in one step.

Pressing [ENTER] brings this message because we have made an error. Press 2 to "goto" the error.

We get this screen which has a blinking cursor on the last parenthesis. This means we have an extra right parenthesis which has no matching left parenthesis.

This screen shows the result when we go back and insert the missing left parenthesis into the calculation. We get the same result as before.

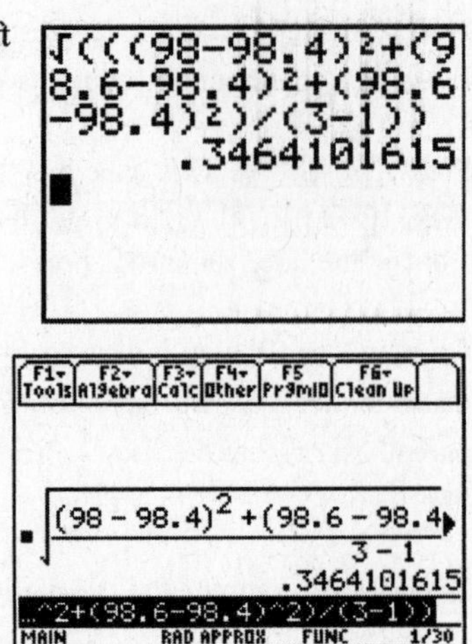

The screen at right shows the same calculation done on an 89 calculator. One important difference here is that the 89 does not have the [x^2] key. To exponentiate to any power, use the [^] key followed by the desired power. Also notice the [▸] at the right of the output display. This is a cue that there is more to be seen. Press the up arrow to highlight the output display, then press to right arrow to scroll to the end.

STORING LISTS OF DATA

Your text, like the real world, is full of sets of real data. In order to analyze the data, you must store it in a list in your calculator. Here, we will show you how to do this. Additionally, you should know that all of the data sets in the Appendix of your textbook are available to be transferred to your calculator from another calculator (or from a computer). You can then use the [APPS] key (on an 83 Plus or 84) to transfer a set into your lists. See your instructor and/or the Appendix of this companion for more information. All of this is analogous using the Nspire calculations app; the "store" function is [ctrl][sto→ var].

Storing Lists of Data from the Home Screen

EXERCISE: Random Sample of 10 Students: Revisited from pages 5 and 6 of this companion: Ten random integers between 1 and 60 were generated and displayed in a list. We will modify the example to store the integers in list L2 on a TI-83 or 84 (list2 on an 89). We will then store the data in a list which we name ourselves.

Modify the command by repeating the first three lines, but this time stipulate that the integers you generate will be stored in L2. Do this by typing [STO▸] [2nd] [2] after the randInt(1,60,10). Press [ENTER] for the screen at right. (Remember, that this was done explicitly in the random number generation on an 89).

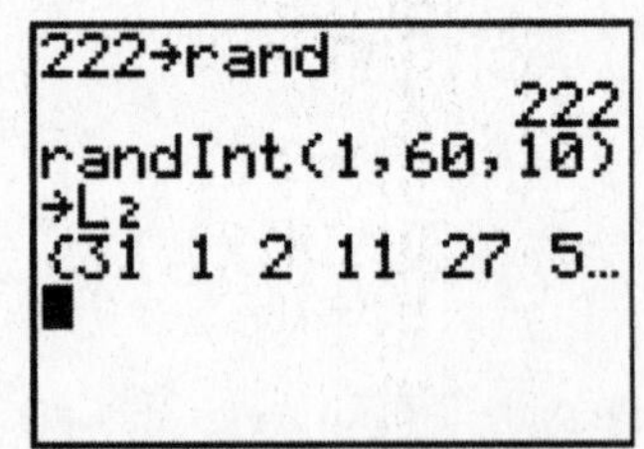

To recall the list you have stored, so that it all appears on the Home screen press [2nd] [STO▸] [2nd] [2]. (The sequence [2nd] [STO▸] chooses the RCL (or recall) function and [2nd] [2] chooses list L2 as what will be recalled.)

To recall a list on the home screen of an 89, we also use the recall ([2nd] [STO▸]) command. However, this brings up an interim screen, which asks what you want to recall.

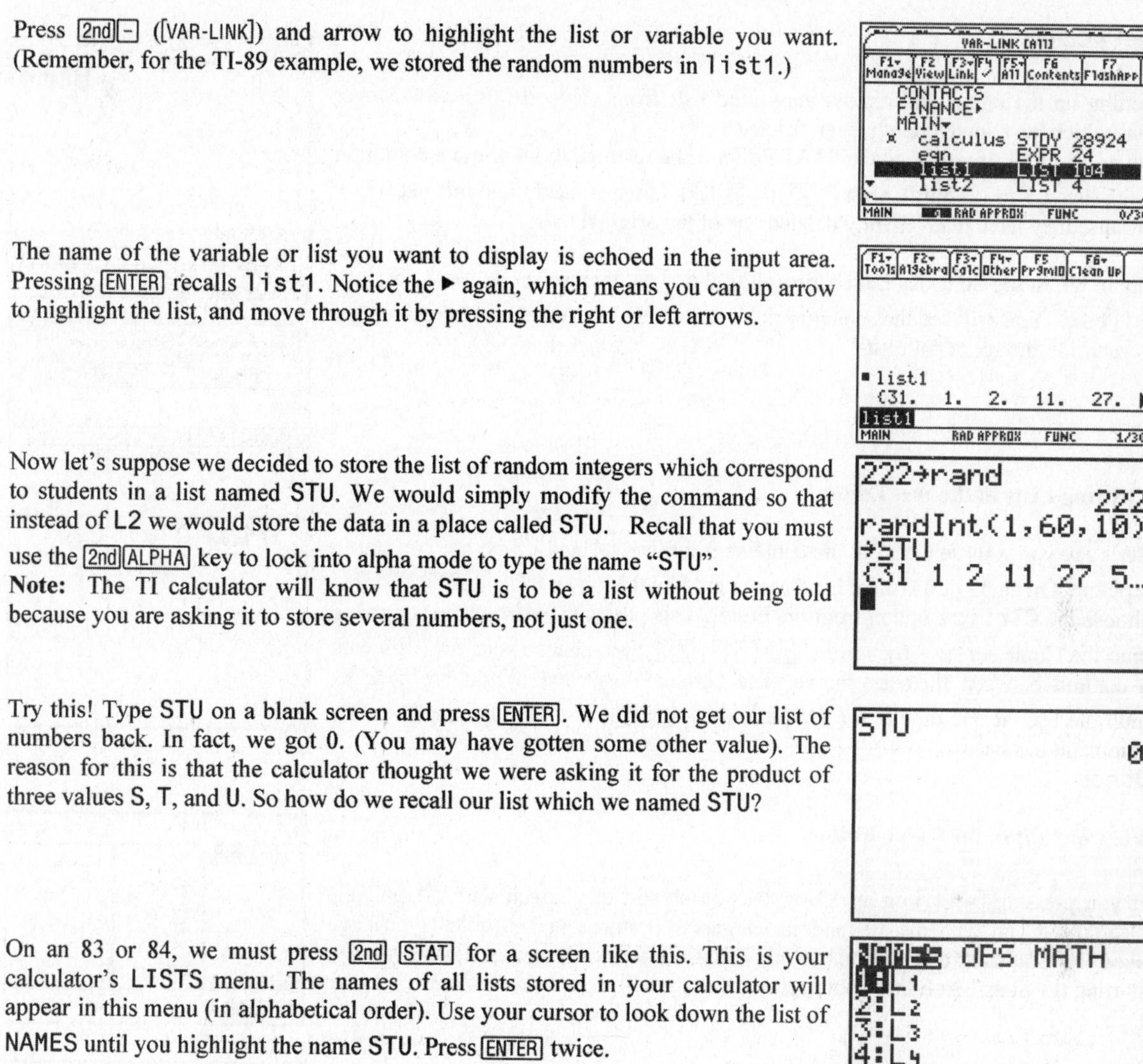

Press [2nd][−] ([VAR-LINK]) and arrow to highlight the list or variable you want. (Remember, for the TI-89 example, we stored the random numbers in `list1`.)

The name of the variable or list you want to display is echoed in the input area. Pressing [ENTER] recalls `list1`. Notice the ▶ again, which means you can up arrow to highlight the list, and move through it by pressing the right or left arrows.

Now let's suppose we decided to store the list of random integers which correspond to students in a list named STU. We would simply modify the command so that instead of L2 we would store the data in a place called STU. Recall that you must use the [2nd][ALPHA] key to lock into alpha mode to type the name "STU".
Note: The TI calculator will know that STU is to be a list without being told because you are asking it to store several numbers, not just one.

Try this! Type STU on a blank screen and press [ENTER]. We did not get our list of numbers back. In fact, we got 0. (You may have gotten some other value). The reason for this is that the calculator thought we were asking it for the product of three values S, T, and U. So how do we recall our list which we named STU?

On an 83 or 84, we must press [2nd] [STAT] for a screen like this. This is your calculator's LISTS menu. The names of all lists stored in your calculator will appear in this menu (in alphabetical order). Use your cursor to look down the list of NAMES until you highlight the name STU. Press [ENTER] twice.

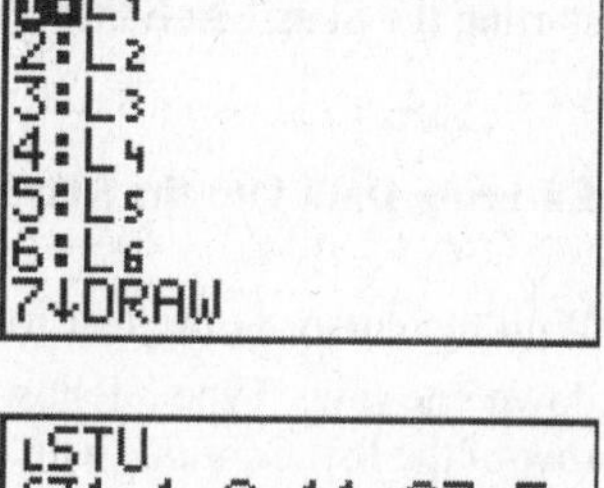

You should see this. You will notice that what was pasted on the Home screen is LSTU and not simply STU. So while we did not need to specify that STU was a list when we first stored our data, we did need to specify this fact when we were trying to recall it. In situations where it is unclear whether or not you need the small L in front of the name of a list, your safest bet is to paste the name in from the LISTS, NAMES menu as we did above.

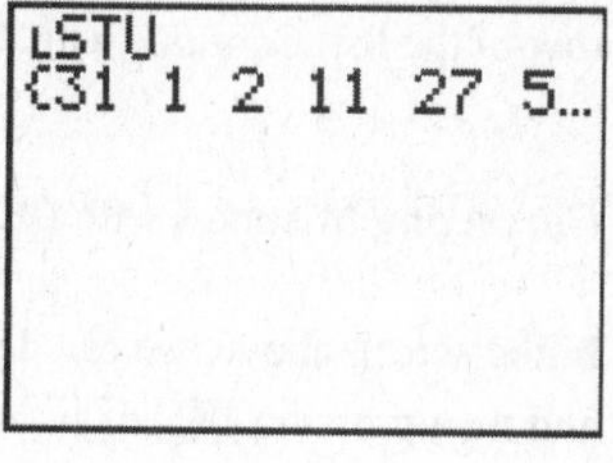

Storing Lists of Data Using the STAT Editor

Using the STAT Editor is the easiest way to store lists and work with the data therein. The STAT Editor comes with 6 lists named L1 through L6 (`list1` through `list6` on an 89). Other lists can be added if desired. The number of lists is only limited by the memory size.

SetUpEditor

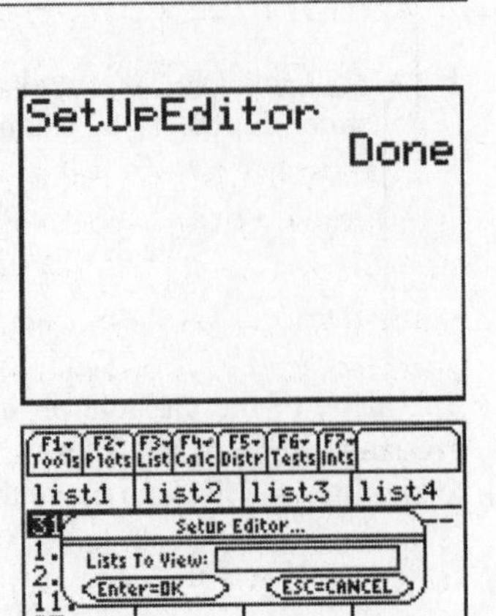

Setting up the editor will remove unwanted lists from view. It also will recover lists which have inadvertently been deleted
On an 83 or 84, if you want the STAT Editor to be restored to its original condition (with lists L1 to L6 only), press [STAT] [5] [ENTER]. Often students find this necessary because they have inadvertently deleted one of the original lists.

On an 89, in the Statistics Editor, press [F1] (Tools), then select option 3:Setup Editor. You will see the screen at right. Leave the box empty and press [ENTER] to return to the six default lists.

Clearing Lists in the Stat Editor

```
ClrList L1,L3
                Done
```

Let's say you want to clear out the contents from lists L1 and L3 before you start a problem. On an 83 or 84 (there is no command like this on an 89), press [STAT] [4] to choose the ClrList option from the Editor. This pastes the start of the command onto the Home screen. Then press [2nd] [1] [,] [2nd] [3] to choose your two lists with a comma between them as shown. The comma is needed if you are clearing multiple lists at one time as we are here. Press [ENTER] to see the message that the command has been
Done.

View or Edit in the STAT Editor:

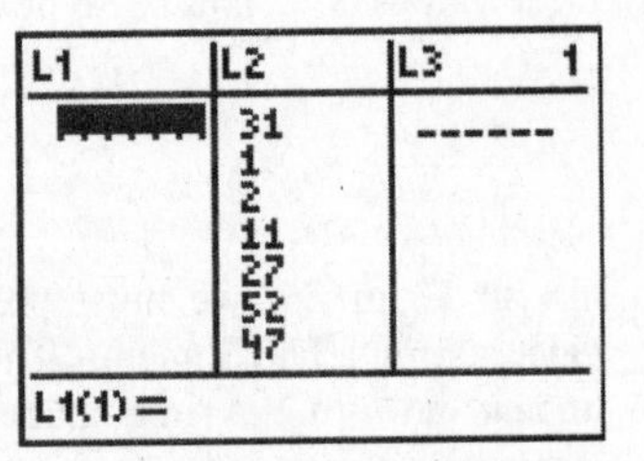

If you press [STAT] [1] on an 83 or 84, you should this screen with L1 and L3 cleared out and with our 10 random integers still stored in L2 (unless you have used the Statistics editor in some other class before). If you are using an 89, starting the Stats/List Editor app displays the editor.

Entering Data into the STAT Editor

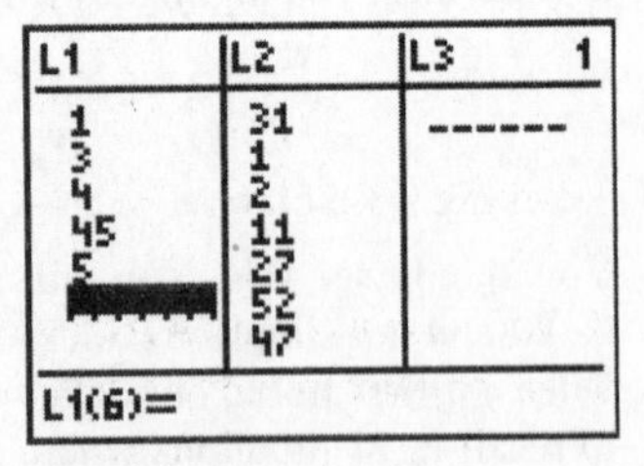

With the cursor at the first row of L1, type 1 and press [ENTER]. The cursor moves down one row. Type 3 followed by [ENTER] and the 3 will be pasted into the second row of the list. Continue with 4, 45, and 5 as seen at right.

Correcting Mistakes with DEL and INS

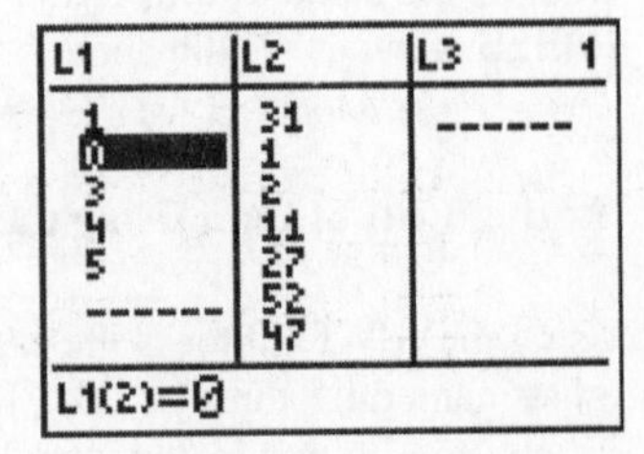

In the screen above, we can delete the 45 by using the [▲] key until it is highlighted and then pressing [DEL] ([♦][←] on an 89). To insert a 2 above the 3 move the cursor to the 3 then press [2nd] [DEL] ([2nd][←] on an 89) (to choose INS or insert mode). Note a 0 was inserted where you wanted the 2 to go.

Clearing Lists without Leaving the STAT Editor

Suppose you wish to clear a list, say L2, while you are still in the STAT Editor. You should use the cursor to highlight the name of the list at the top. With the name highlighted press [CLEAR] and you will see this. Press [ENTER] and the contents of the list will be cleared. *Make sure not to press* [DEL] or the list will be deleted entirely and you will have to use SetUpEditor as described above to retrieve it.

Storing Data with a Named List

Suppose we wish to store the 10 random numbers 17, 44, 43, 28, 27, 51, 30, 39, 34, and 32 in a list named Rand2. This procedure is the same on all calculators.

First with L2 highlighted at the top press [2nd] [DEL]. L2 will move to the right and there will be a new list inserted in its place. The calculator will be in ALPHA mode and be prompting you for the name of your new list.

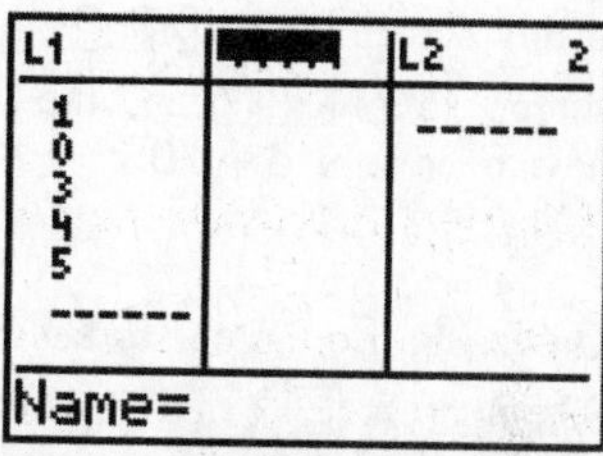

Type RAND and then press [ALPHA] to release the calculator from ALPHA mode, so you can type the final character in the name which is 2.

Press [ENTER] and then [▼] (no need to press ⊙ on an 89). Now input the desired values into the list as at right.

Note: If the list Rand2 had already been created, its name could have been pasted into list name. In either case, if the list already existed its entire contents (if there were any) would have been pasted in as well as its name.

Deleting a List from the STAT Editor:

If you wish to delete a list from your STAT Editor, simply highlight the list name and press [DEL]. The name and the data are gone from the Editor but not from the memory.

Using SetUp Editor to Name a List:

On a TI-83 or -84 home screen, press [STAT] [5] to call SetUpEditor. Now type IDS, STU, L1. You will have to be careful to keep pressing the [ALPHA] key before each character or release the alpha lock ([2nd][ALPHA]) to type the commas. Then press [ENTER]. Then press [STAT] [1] to view your lists.

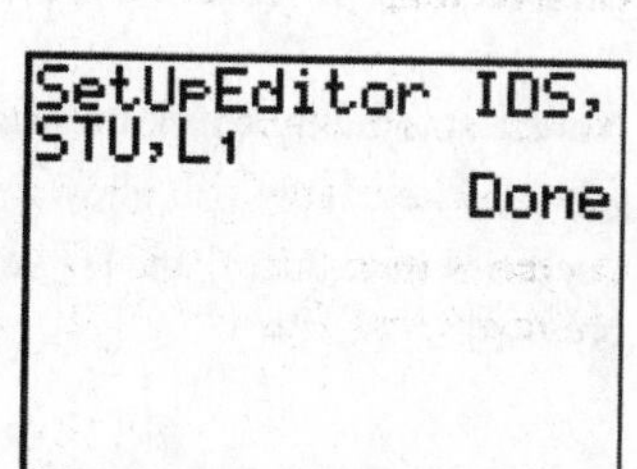

In the screen at right, you will see the old list STU has been placed back in full view. Also there is a new list IDS ready and waiting for some data to be entered. Try pressing the right arrow to see more lists – there aren't any. This command set the editor with just the three lists specified.

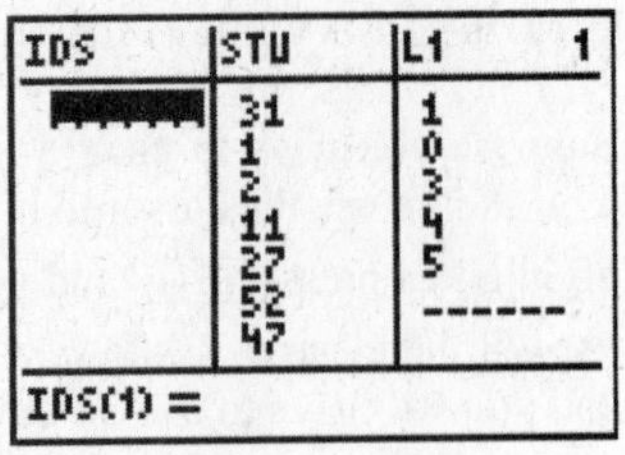

To do this on an 89, follow the steps outlined above to generally set up the editor, and type in the list names just as one does for the 83 or 84.

Making a Copy of a List

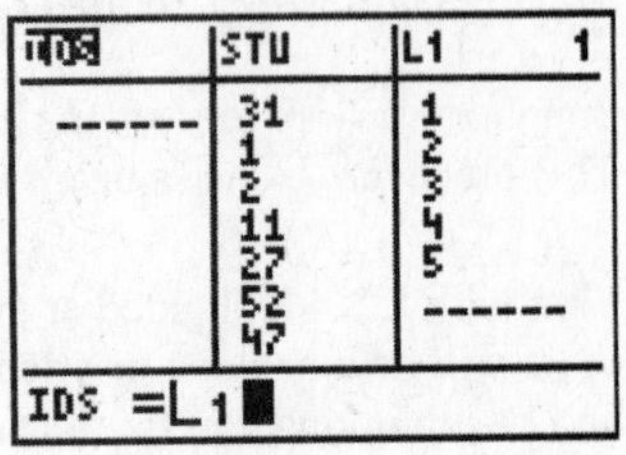

On an 83 or 84, use cursor control keys to highlight the top of the new list IDS as in my screen. Press 2nd 1 to paste in the name L1 as in the bottom line of the screen. Press ENTER and the L1 data will appear under IDS. The contents of L1 have been copied to IDS.

On an 89, use the cursor keys to highlight the top of the new list ids as with the 83,but press F3 (list). Now, press ENTER to select option 1:Names. Scroll down to find the name of the list you wish to copy, and press ENTER to select it. The list name will be pasted in as at right. Pressing ENTER again will copy the list.

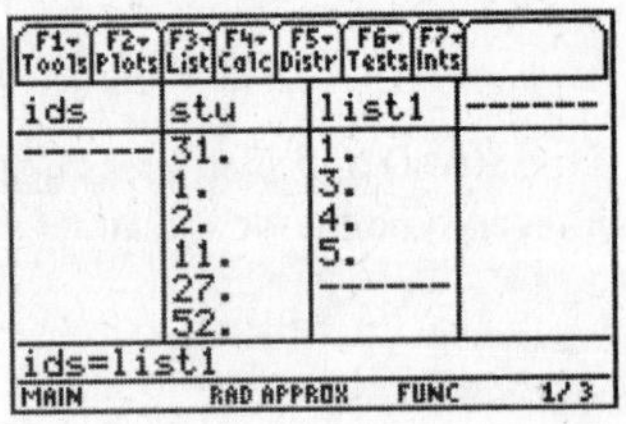

Generating a Sequence of Numbers in a List

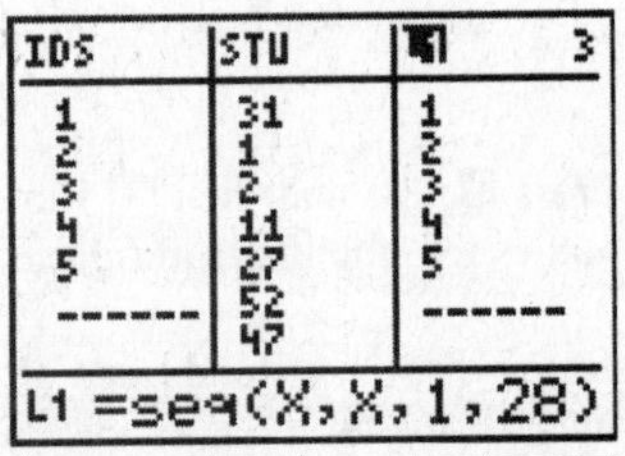

From time to time one may want to enter a list of sequenced values (years for example in making a time-series plot). It is certainly possible (but tedious) to enter the entire sequence just as one would enter normal data. There is an easier option, however.

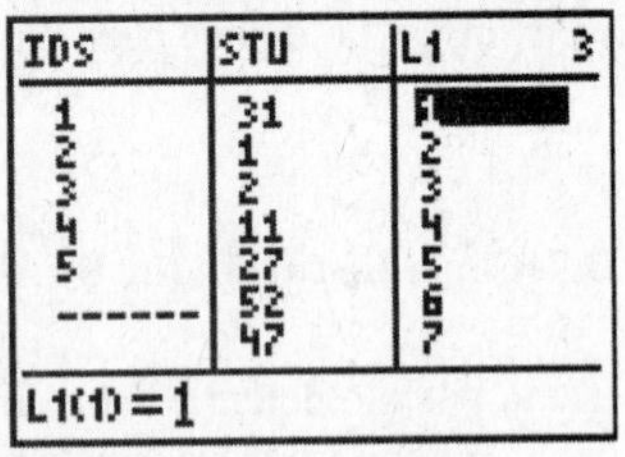

Use cursor control keys to highlight L1 in the top line. Press 2nd STAT ▸ 5. You are choosing the LIST menu and then choosing the OPS submenu. From the OPS submenu you are choosing option 5 which is seq(. This has been pasted onto the bottom line of the screen. Type in the rest so that you have seq(X,X,1,28. Press ENTER and the sequence of integers from 1 to 28 will be pasted into L1 as in my screen.

On an 89, the procedure is analogous, but access the LIST OPS menu by pressing F3 2, then select option 5.

Note: To quickly check the values on a multi-screen list you can press the green ALPHA key followed by either the ▴ or ▾ key. This will allow you to jump up or down from one page (screen) to another. The green arrows on the keyboard near the ▴ and ▾ keys are there to remind you of this capability. On a TI-89, instead of the ALPHA key, press ♦.

DELETING LISTS

Deleting a Named List from Memory (TI-83/84)

To remove both the name of a list and its data from RAM press [2nd] [+] to call the `MEM` function. You should see this screen.

```
MEMORY
1:About
2:Mem Mgmt/Del...
3:Clear Entries
4:ClrAllLists
5:Archive
6:UnArchive
7↓Reset...
```

Press [2] to choose `Mem Mgmt/Del`. You will get a screen like this.

Press [4] to choose to see a list of *List* names. Use [▼] to move down the display to the list you want to remove (say `IDS`) as shown in this screen.

```
RAM FREE  21672
ARC FREE 298628
  L5          12
  L6          93
  IDList      21
  DRAW        42
▸ IDS         59
  N1          40
```

Press [DEL] to delete the list. You can remove lists one by one from this screen.

Press [2nd] [MODE] to `QUIT` and return to the Home screen.

Note: (For TI-83 Plus and TI-84 only) If the list was also saved in a Group stored in Archive memory, the above procedure has not deleted the copy. (See the Appendix for more information on Groups.) Groups can be deleted from the calculator if `4:List...` with `8:Group...`.

Deleting a Named List from Memory (TI-89):

In the Statistics/List Editor, Press [F3] (`Lists`) and press [ENTER] to select option `1:Names`.

```
F1 Tools | F2 Plots | F3 List | F4 Calc | F5 Distr | F6 Tests | F7 Ints
1:Names...
2:Ops ▸
3:Math ▸
4:Attach List Formula...
5:Delete Item        ♦DEL
21.
22.
23.
list1[20]=20.
MAIN   RAD APPROX   FUNC
```

Move the cursor to highlight the name of the list you want to delete. Now, press [F1] (`Manage`). Option 1 on this submenu is `Delete`. Press [ENTER] to select it.

```
VAR-LINK [All]
F1 Manage | F2 View | F3 Link | F4 ✓ | F5 All | F6 Contents | F7 FlashApp
CONTACTS
FINANCE▸
MAIN▾
  ids      LIST 44
  list1    LIST 284
  list2    LIST 4
  list3    LIST 4
MAIN   RAD APPROX   FUNC   3/3
```

You will get the screen at the right which asks you to confirm the deletion. Pressing [ENTER] will complete the delete, [ESC] will cancel the operation.

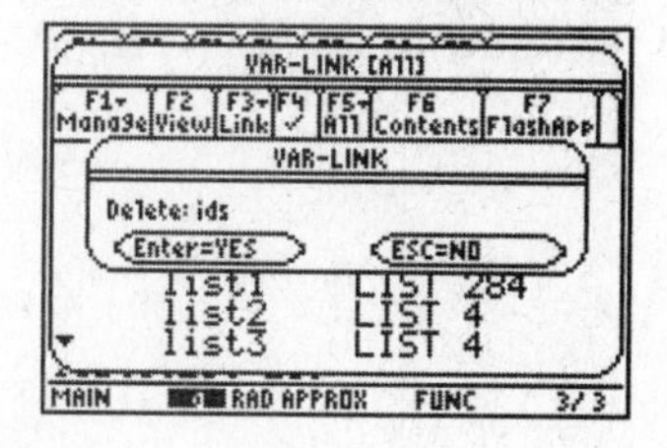

Using the Augment Function for Large Lists

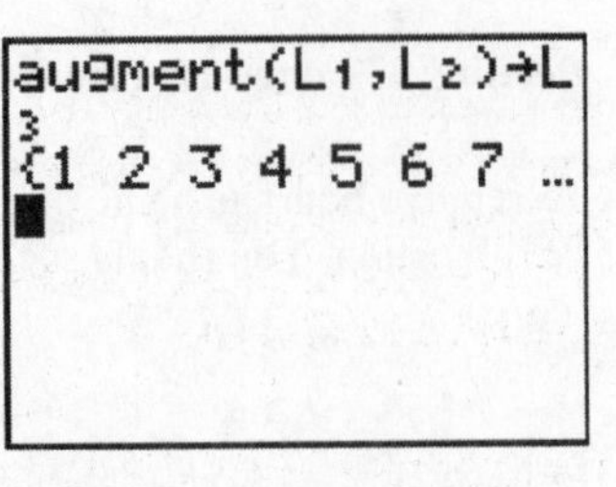

Several lists can be combined into one large list by using an option which is on the LIST, OPS submenu at option 9. This is the augment option. For example if you wish to combine lists L1 and L2 and store the combined list in L3, you can press [2nd] [STAT] [▶][9] to paste the augment(on your Home screen. Then put the names of the lists you wish to combine on the screen separated by commas. Close the parentheses. Then press [STO▶][2nd] [3] to store in L3. Finally press [ENTER]. This function will only work if there is actually data in the lists you are augmenting.

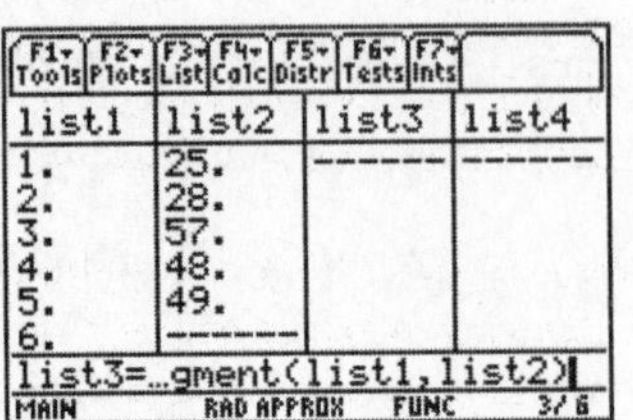

To do this procedure on an 89, place the cursor at the top of an empty list to highlight the list name, then press [F3] (List). Press [2] to select the Ops menu, where the augment function is choice 8. Press [8] to paste it into the input area. Now press [F3] again followed by [ENTER] to select the names list. Arrow to the names of the lists you want to combine and press [ENTER] to paste their names into the command. Be sure to separate each name with commas and finish the command by closing the parentheses.

Note: The "augment" function could be used so that several students could cooperate in typing in a large set of data. Each student could type part of the data set into a different list (L1, L2, L3, etc.) Then the calculators could be linked, the lists transferred to one calculator and then the augment function could be used to combine them into one large list which could then be redistributed for all to share.

WORKING WITH LISTS (Nspire)

Using the Nspire, lists are really columns in the spreadsheet application. In order to use the "list" for statistical graphs, it must be named by entering whatever name you wish in the top line of the list. The second entry box is used for entering formulas (always preceded with ⊜ at the upper left of the keyboard).

This calculator is not well equipped for deleting lists; you can clear entries by pressing [shift] and using the arrow keys to highlight a group of cells; pressing [clear] will clear all highlighted cells. In many instances it is easier to add a new "spreadsheet" to the problem (press [home] [3]) or create a new "problem" by first saving the current (press [home][7], then [menu] and select [3] Save As) or deleting the current "problem" by pressing [home][7], then [clear]. You will be asked to verify the delete before it actually takes place.

WHAT CAN GO WRONG?

Why is my list missing?
By far, the most common error, aside from typographical errors, is improper deletion of lists. When lists seem to be "missing" the user has pressed [DEL] rather than [CLEAR] in attempting to erase a list. Believe it or not, the data and the list are still in memory. To reclaim the missing list press [STAT]and select choice 5:SetUpEditor followed by [ENTER] to execute the command. Upon return to the editor, the list will be restored.

2 Summarizing and Graphing Data

This chapter introduces the graphical plotting and summary statistics capabilities of TI calculators. These calculators can display histograms, scatter (*X-Y*) plots, connected scatter plots (used for time-series graphs, frequency polygons and ogives), normal plots (used to determine whether a data set is approximately bell-shaped), and boxplots. The TI-Nspire can also create dotplots, pie charts and bar charts. This chapter will deal only with those plots described in Chapter 2 of the text.

The [2nd] [Y=] (STAT PLOTS) menu on the TI-83 and -84 are used to obtain descriptive plots of data sets. On the TI-89, plots must first be defined and then displayed. As procedures differ, we will first describe how to create a plot using TI-83/84 calculators, then repeat as necessary for the -89 series as well as the Nspire. The author assumes you have read and familiarized yourself with the content of Chapter 1.

We will use the data on pulse rates presented in Table 2-1 of the text to illustrate first the procedure for creating a histogram from raw data. This is by far the most common means of displaying data using technology. Procedures for using data which have already been summarized into a table will be presented later. The table is repeated below for convenience.

Pulse Rates of Females									
76	72	88	60	72	68	80	64	68	68
80	76	68	72	96	72	68	72	64	80
64	80	76	76	76	80	104	88	60	76
72	72	88	80	60	72	88	88	124	64

Enter the pulse rates into list L1 (TI-83/84) or list1 (TI-89).

One preparatory step which should be undertaken before defining any statistics plot is to ensure that all functions entered on the [Y=] screen have been erased; if not, these will also display and may also cause dimensioning problems for the calculator. Press the [Y=] key on the upper right on a TI-83/84, or go to the Y= app on a TI-89 and ensure that all functions are cleared (Press [CLEAR]) as at right.

Plot1 Plot2 Plot3
\Y1=
\Y2=
\Y3=
\Y4=
\Y5=
\Y6=
\Y7=

CREATING A HISTOGRAM – TI-83/84

The next step is to define the plot. This is done by pressing [2nd][Y=] (Stat Plot). You will see the screen at right. Notice that there are three plots which can be displayed at any one time. For most purposes, there should be only one turned "on" at once. Notice here Plot1 is On and Plots 2 and 3 show Off. Scrolling down the menu are options 4 and 5 that turn all plots off or on with a single command. Selecting these will transfer the command to the home screen. Executing it requires pressing [ENTER].

STAT PLOTS
1:Plot1...On
L1 1
2:Plot2...Off
L4 1
3:Plot3...Off
L3 1
4↓PlotsOff

Press [ENTER] to select Plot1. The cursor should be blinking over the word On. If On is not already highlighted, press the right or left arrow key to move the blinking cursor and [ENTER] to move the highlight and to select displaying the plot. Note there are six graphics types. Histograms are the third choice. Pressing [▼] will move the cursor to the first plot type. Use [▶] to move the cursor to the histogram figure. Press [ENTER] to actually move the highlight.

Plot1 Plot2 Plot3
On Off
Type:
Xlist:L1
Freq:1

At this point, your screen probably looks like the one on the previous page. We're ready to display the graph, since our data was in list L1 and each data value had frequency 1 (representing one occurrence of the value.) If you want to graph data in other lists, move the cursor to Xlist: and enter the list name (2nd n, where n is the number of the list). We'll talk more about frequencies later. Notice if you move the cursor to Freq: it will flash as A. If you need to change this back from something else to a 1 you will need to press ALPHA before typing the 1.

The easiest way to display a histogram (or any statistics plot) is to press ZOOM 9 (Zoom Stat). The resulting graph is seen at right. The *X*-axis "floats" a little way up from the bottom of the screen. This is so that values as seen in the next picture do not interfere with the plot.

You will want to see exactly what the graph shows. To do this, press TRACE. A blinking cursor will appear in the first bar at the left of the graph. At the bottom of the screen the minimum value included in the bar, maximum value for the bar, and number of observations in the bar will be displayed. This bar goes from 60 to 70.666667. There are 12 observations in this interval, indicated by n=12. Pressing the right arrow key (▶) will allow you to continue through the graph seeing the interval ranges and numbers of observations in each bar.

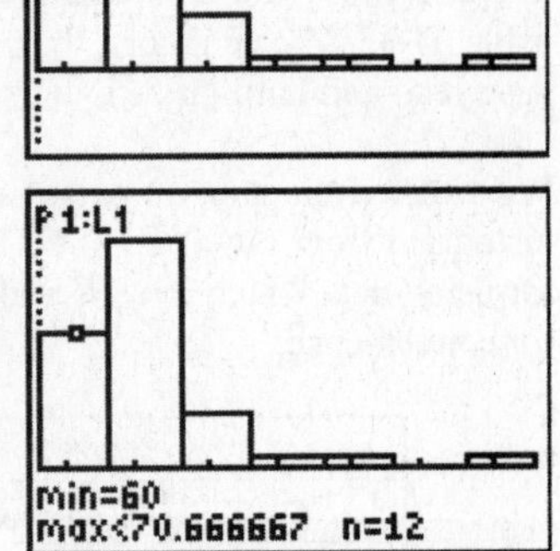

So far, we can see the distribution of women's pulse rates appears to be unimodal and right skewed, as the highest bar is toward the left and frequencies drop consistently as we move to the right. Visually, the center of the graph (the midpoint) will be somewhere around 75 beats per minute.

There is a downfall to using simply ZOOM 9 for histograms. Look again at the first interval. It doesn't really make sense in a natural way. The bar width represents a difference between the low and high ends of each bar of 10.66667... which is unnatural. We'd like to fix this.

Manipulating Windows on the TI-83/84

To force particular minimums, maximums and scaling we will press WINDOW. This displays the screen at right. Notice the Xmin was the smallest value shown on the plot and Xscl was the bar width. These are what we'd like to change. You generally won't have to change any of the *Y* variables here (unless you lose the top of a bar – then increase Ymax).

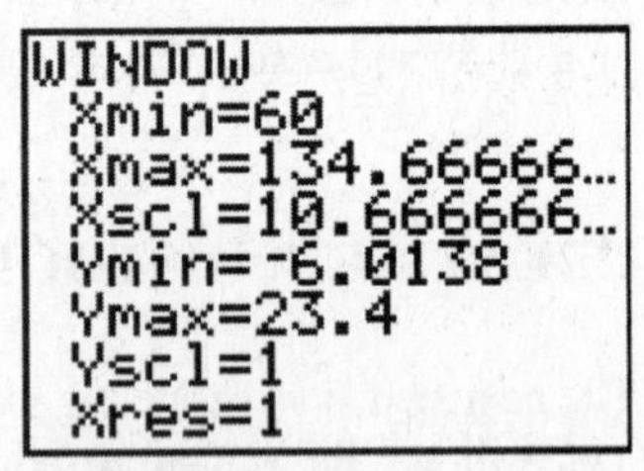

Since 60 is a logical starting point for the graph, we won't change that but we will change Xscl to 10 (This sounds pretty reasonable and will duplicate the frequency distributions used in the text.) We also changed Ymax to 16 so that the graph will use most of the display screen.

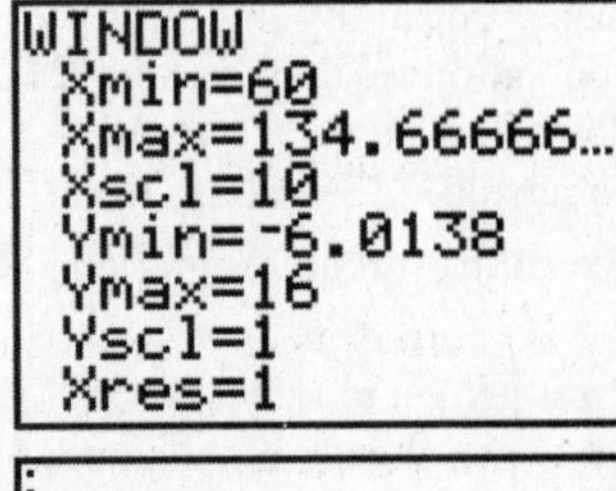

To display the new graph, press GRAPH. *NEVER* press ZOOM 9 after changing a window. You'll just go back to the one you had before! Notice that the first bar includes pulse rates from 60 to less than 70 (or at most 69.999). This graph replicates the histogram one would get using the tabulated data as in Table 2-2 of the text, from which the histogram in Figure 2-3 was obtained.

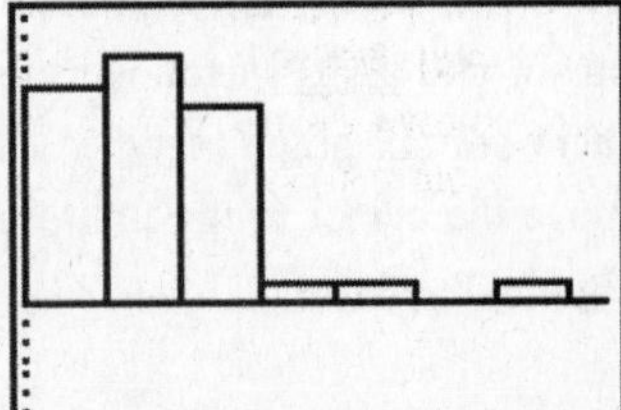

CREATING A HISTOGRAM – TI-89

Once data have been entered into a statistics list, the first step is to define the plot. This is done from the Statistics Editor by pressing [F2] (`Plots`) followed by [ENTER] to select `Plot Setup`. You will see the screen at right. Notice that there are nine plots which can be displayed at any one time. For most purposes, there should be only one active (checked) plot at once.

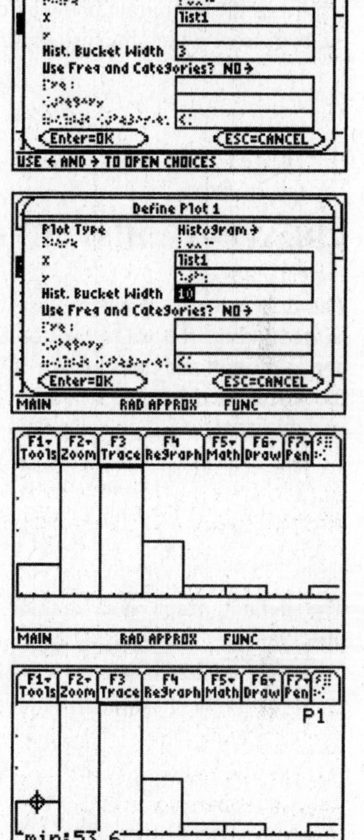

Press [F1] to select defining `Plot1` since it was highlighted. The cursor should be blinking over the plot type. If not already set to a histogram, pressing the right arrow gives a menu of five plot types. Move the cursor to highlight choice `4:Histogram` and press [ENTER] to select it or press [4].

Press the down arrow to the box labeled `x`. Press [2nd][−] (`VAR-LINK`) to get the list of list names. Move the cursor to highlight the one you wish to use, press [ENTER] to select it. The TI-89 then wants the histogram bucket width which is the bar width. Press the down arrow to move to this box. This is something that may have to be "played around with" to get a good picture. Here, I have set the bar width to 10. Press [ENTER] to complete the plot definition. You will be returned to the `Plot Setup` menu.

The easiest way to start displaying a histogram (or any statistics plot) is to press [F5] (`Zoom Data`). The resulting graph is seen at right. The x-axis "floats" a little way up from the bottom of the screen. This is so that values, as seen in the next picture, do not interfere with the plot. This picture shows one bad point of the TI-89. It does not always get the windowing correct for histograms. We don't see the full heights of the second bar. We'll change that later.

You will want to see exactly what the graph shows. To do this, press [F3] (`Trace`). A blinking cursor will show in the first bar at the left of the graph. At the bottom of the screen the minimum value included in the bar, maximum value for the bar, and number of observations in the bar will be displayed. This bar goes from 53.6 to 63.6. There are seven observations in this interval, as indicated by the `n:3` at the bottom right. Pressing the right arrow key (▶) will allow you to continue through the graph seeing the interval ranges and numbers of observations in each interval.

At this point, we can see the distribution of women's pulse rates appears to be unimodal and relatively right-skewed as distribution is much longer on the right-hand side of the peak. We see the center is around 75 beats per minute. We'd really like to see the whole graph, however.

There is another downfall to using simply `Zoom Data` for histograms. Look at the intervals. They really do not make sense in a natural way. We'd like to fix this.

Manipulating Windows on the TI-89

To force particular minimums, maximums and scaling we will press [◆][F2] (`Window`). This displays the screen at right. Notice the `xmin` was the smallest value shown on the plot; `xmax` is the largest. `ymin` and `ymax` are analogous. `xscl` and `yscl` are the distances between axis "tick marks".

```
xmin=53.6
xmax=130.4
xscl=3.2
ymin=-3.
ymax=12.
yscl=3.
xres=1.
MAIN    RAD APPROX    FUNC
```

Change `xmin` to 60 (the smallest data value) `ymin` to –3 and `ymax` to 16. Why must `ymax` be so large? The "tabs" at the top of the screen will hide some of the plot unless it is sized on the roomy side. `Ymin` is set low so that the legends which appear after pressing `Trace` don't obscure portions of the graph.

```
xmin=60.
xmax=130.4
xscl=3.2
ymin=-3.
ymax=16.
yscl=3.
xres=1.
```

To display the new graph, press [♦][F3] (`Graph`). This looks better, and also replicates the histogram one would get using the tabulated data as in Table 2-1 of the text, from which the histogram in Figure 2-3 was obtained.

CREATING A HISTOGRAM (AND DOTPLOT) – TI-Nspire

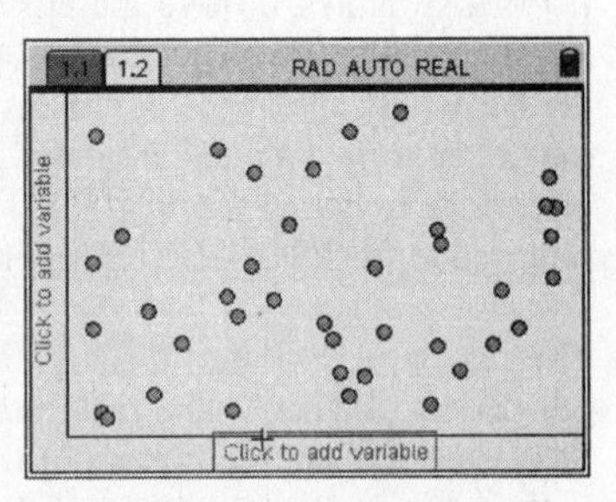

The data have been entered into the spreadsheet in column A which has been named pulse. Return to the home screen and press 5 for the Data and Statistics app. You will see the screen at right. Use the arrow keys to move the cursor to the lower area labeled Click to add variable. Press the grab key for a list of variables; select the one you want and press the grab key again.

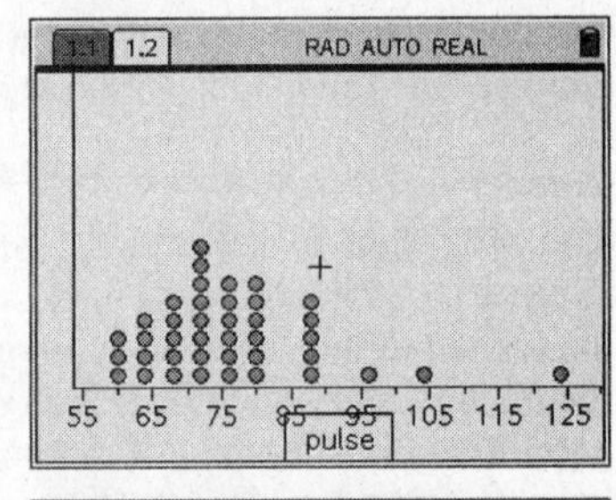

The default graph is a dotplot as seen at right. Each dot represents one observation. To change this to a histogram, press the menu key, then select `1:Plot Type`. Histograms are type 3.

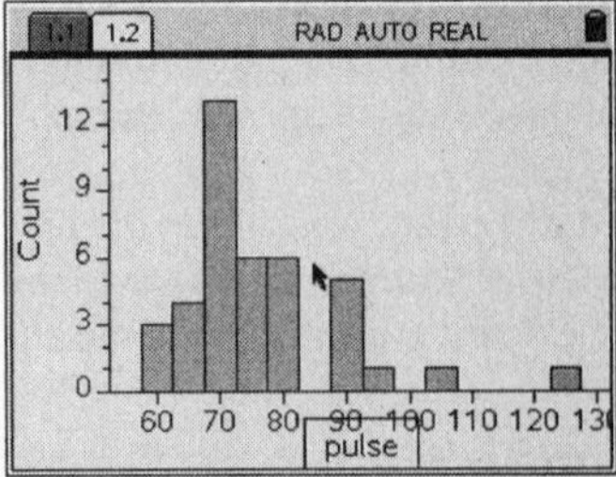

Here is the default histogram. Note that the Nspire makes its tick marks in the middle of intervals. We appear to have bars of width 5 and the graph starts slightly below 60. Let's change that to match the other calculators. Press the menu key, arrow to `2:Plot Properties`; option `2:Histogram Properties` should be highlighted. Press the right arrow to expand the options and select `2:Bin Settings`.

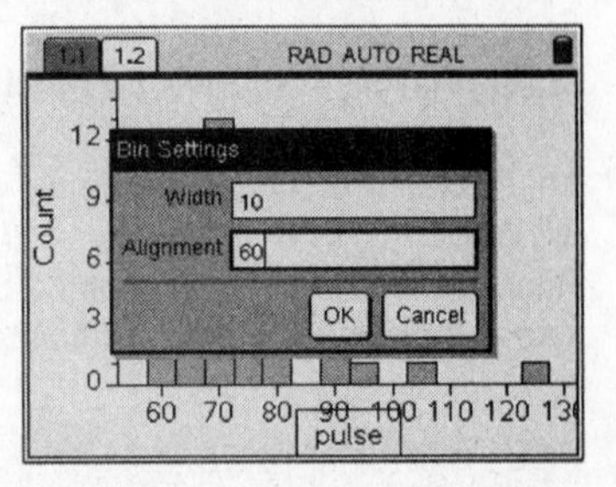

We've selected a bar width of 10 and starting point (alignment) 60. Press [ENTER] for OK.

The finished histogram is shown at right.

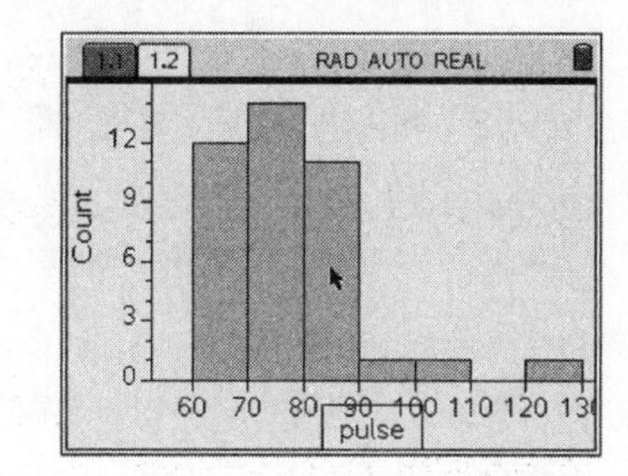

CHANGING THE NUMBER OF BARS

Histograms are a subjective type of plot. How many bars to display (and what the lowest display value for the variable) can be manipulated at will until one has a "nice" plot. Your instructor may give you some guidelines. One "rule of thumb" for many years was to have somewhere between 5 and 20 bars; for most smaller data sets dividing the number of observations by 5 gives a good estimate of how many bars will give a decent picture.

On TI-83/84 calculators, the bar width is controlled using `Xscl` on the window settings screen. On theTI-89 series, one changes the bucket width on the plot setup screen. As seen above, with the Nspire, use `Bin Settings` from the `Plot Properties` menu.

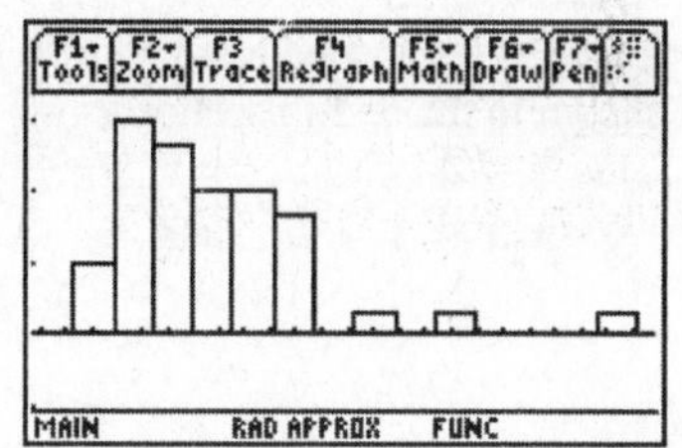

The screen at right was created using an `Xmin` of 60 and a bucket width of 5. `Ymax` was reset to 10 for picture resolution. Just as having too few intervals can hide the true nature of a distribution, having too many can also distort the picture. But here again, we see the unimodal, right-skewed distribution of the previous histogram.

"Printing" the Picture

Unfortunately, calculators do not have printers. To make a hard copy of the graph once you are satisfied with it, use the TRACE key to examine the entire graph. Draw a picture of the histogram, clearly labeling each axis and giving the graph a title. Remember that the intervals given are the endpoints of the intervals. Label them as such. When you are finished, you should have a picture like the one at right. If you have TI-Connect software, you can use the screen capture application to save the picture on the computer for printing directly from the application, or include the graph as a part of a word processing document.

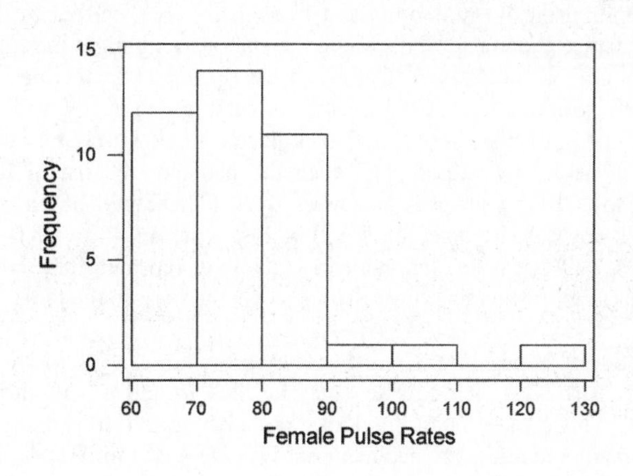

HISTOGRAMS FOR TABULATED DATA – TI-83/84/89

When data are given in a tabulated list, such as in Table 2-5 which gives the IQ scores of a sample of 1000 adults, we can still create a visual histogram of the data. The data are reproduced at right for convenience.

IQ Score	Frequency
50-69	24
70-89	228
90-109	490
110-129	232
130-149	26

These data are presented in 20 point intervals. We will enter the midpoint of each interval (60, 80, etc) into a list (`L1` or `list1`) and the frequencies into a second list (`L2` or `list2`).

On the TI-83 or 84, we define the plot as has been done before, but instead of using `Freq:1` we specify the list containing the frequencies (`L2`). If using a TI-89, on the plot definition screen, use a bucket width of 2, and use the arrow key to change the answer to `Use Freq and Categories?` from No to Yes, then use `Var-Link` ([2nd][-]) to select the list with the frequencies.

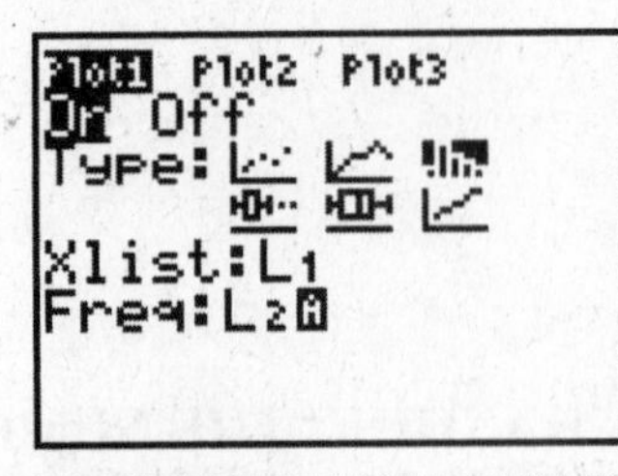

Since we know the bar widths we need (20 points) and the minimum and maximum data values, as well as the frequencies, it is easier in this case to specify the window settings as at right. Set the window for the TI-89 similarly, but remember to give extra height to the y-axis for room for the tabs. To display the histogram, press [GRAPH].

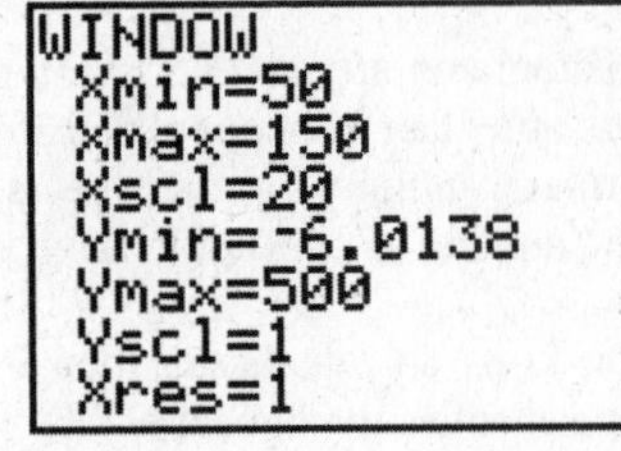

Here is the finished histogram. Compare it to the graphic produced by STATDISK shown in the text in section 2-3.

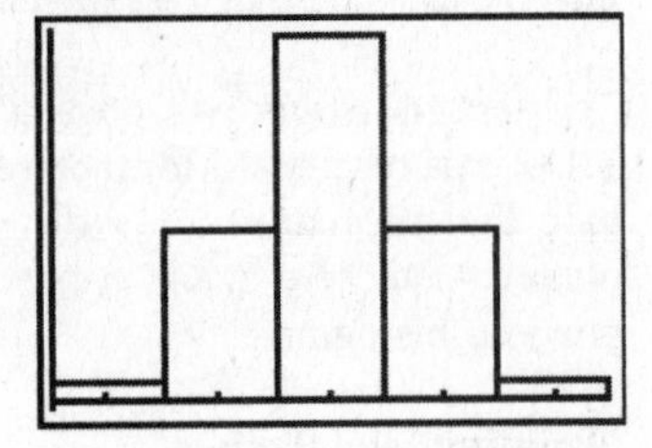

HISTOGRAMS FOR TABULATED DATA – Nspire

We have entered the IQ midpoints and frequencies as specified in the section above, and named the iq and freq. Move the cursor to highlight the function box in an empty column of the spreadsheet. Press (=), then (enter) for the `Catalog`, then press (F) to move to that portion of the catalog. Press (ctrl)(3) to move a screen at a time down the list of functions to locate `freqTable▸list(`. Enter the name of the list of values (`iq`), then the name of the list of frequencies (`freq`). Close the parentheses and press (enter). (There will be an intervening message that any data in the column will be lost – press (enter) for `OK`). You have just created a new list with each value in variable iq repeated freq times. Name this new variable iqs (or something else — so you can create the plot).

Add a new Statistics page to the worksheet by pressing (home)(5). Move the cursor to the lower part of the window so you can click to add a new variable. Add variable iqs. This creates a plot that looks like a bar chart (but is really a dot plot – there are many repetitions of each value). Press (menu), (enter) for `Plot Type`, and (3) for `Histogram`.

So far, we have this histogram, which doesn't look like a real histogram since the bars are not connected. Press (menu), (2) for `Plot Properties`, and (enter) for `Histogram Properties`, then (2) for `Bin Settings`. We'll change `Bar Width` to 20 and `Alignment` (the minimum value) to 50. Move the cursor to highlight OK, then press either (enter) or (click) to display the new plot.

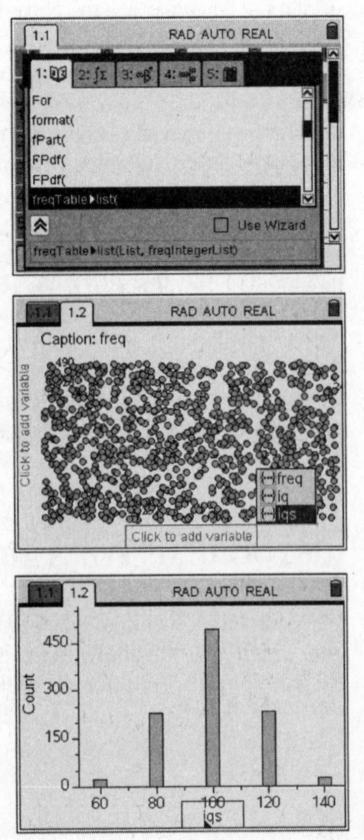

At right is our finished plot. Notice that with this calculator, we have a pretty much "finished" plot – variables and values are nicely labeled.

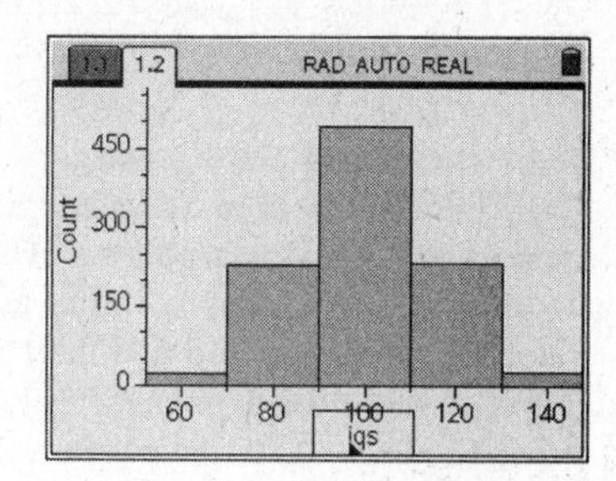

FREQUENCY POLYGONS

A frequency polygon is another type of plot which is used to picture a distribution. It uses line segments to join points that are located at the class midpoint on the *X*-axis and the frequency of the interval on the *Y*-axis. This type of plot is useful for comparing two distributions, as well. We return to the data on pulse rates, reproducing for convenience the data in tabular form. I have labeled the rate column as extending to less than the beginning of the next interval to emphasize that these are, for practical purposes, 10 year intervals. This frequency chart clearly indicates that males tend to have lower pulse rates than females – a fact hidden in the long list of numbers in Table 2-1.

Pulse rate	Female Frequency	Male Frequency
50 - <60	0	6
60 - <70	12	17
70 - <80	14	8
80 - <90	11	8
90 - < 100	1	1
100 - <110	1	0
110 - < 120	0	0
120 - <130	1	0

The midpoint of each interval is the average of the two ends, so the midpoint of each interval ends in 5. When entering the data in the calculator, we use one list for the age midpoints, and two lists for the frequencies. We will also insert two additional intervals (midpoints of 45 and 135) both with frequencies of 0. At right is the first screen of data.

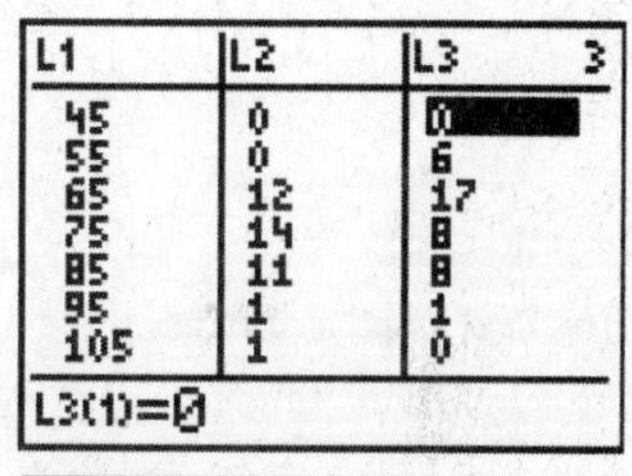

L1	L2	L3
45	0	0
55	0	6
65	12	17
75	14	8
85	11	8
95	1	1
105	1	0

L3(1)=0

To display both distributions at the same time, we will define two graphs to display together. On the TI-83 or 84 we will use the second plot type (⊵). On the 89, use the arrow keys on the plot definition screen to select plot type `2:xyline`. The `Xlist(x)` is the one with the IQ midpoints (L1 or `list1`) and the Ylist (y) is the one with the female frequencies. Choose a mark to display the actual data values. My TI-84 plot definition is at right.

Now we will define a second plot similar to that just done, but using the frequencies (Ylist) for the males. Select a different type of mark for this plot to help distinguish the two. My definition of the second plot on the TI-89 is at right.

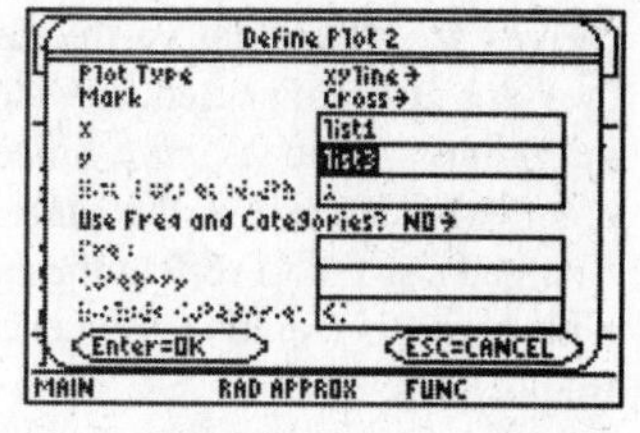

On both calculators, `Zoom Data` (ZOOM 9 or F5) will display the graph. The graph with the square points is for women, the crosses are for the male data. We clearly see again a difference in the pulse rate distributions – the men's distribution peaks about 10 beats per minute lower than the females.

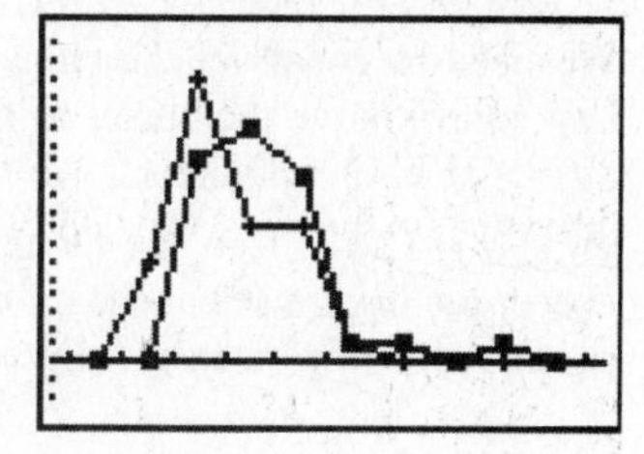

Frequency Polygons – Nspire

We have entered the IQ category midpoints and frequencies as above. Press (home)(5) to add a new Statistics plot page into the "problem." You will see random dots for "data points." Use the arrow keys to move the cursor to the bottom edge and click to add iq as the variable. You will see one dot at each category midpoint. To add the frequencies, you could move the cursor to the left edge and repeat, or press (menu), (2) for `Plot Properties`, then (6) for `Add Y variable`. Select the variable females for the Y variable.

iq	females	males
45	0	0
55	0	6
65	12	17
75	14	8
85	11	8

At this point, you should have the screen at right. We need to connect the dots. Press (menu), (1) for `Plot Type`, then (enter) as option `6:XY Line Plot` will be highlighted. This connects the dots with line segments. To add the males into the plot, press (menu), (2) for `Plot Properties`, then (6) for `Add Y Variable`. Select to enter the males.

The resulting graph is at right. Note that we are given a legend indicating the symbol shape for each gender. To move the legend box, move the cursor inside the box, then hold the grabber (grab) key and use the arrows to move the box so it does not obscure a data point.

The final graph is at right.

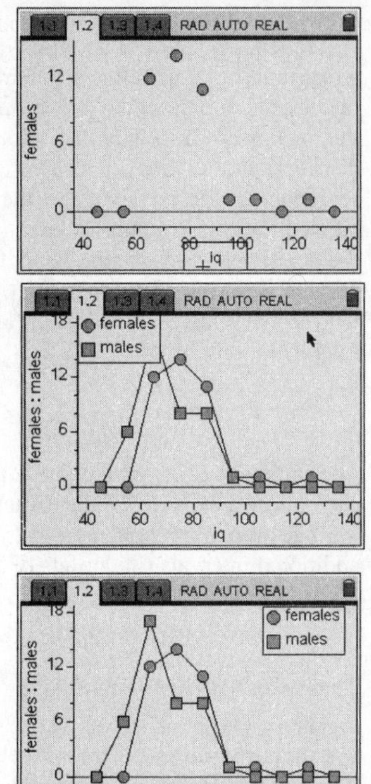

Ogives

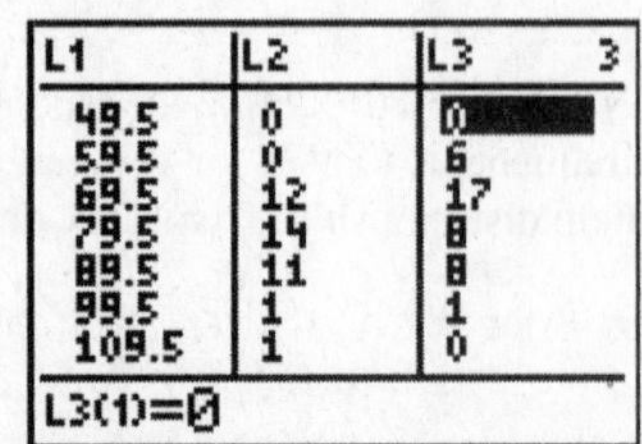

L1	L2	L3
49.5	0	0
59.5	0	6
69.5	12	17
79.5	14	8
89.5	11	8
99.5	1	1
109.5	1	0

L3(1)=0

Ogives are a cumulative frequency polygon. Like the frequency polygons above, they are also connected line plots, but they use the category end values instead of midpoints on the *X*-axis. Since the lowest pulse rate for the females was 60, and practically speaking, the largest age in each category is 69.5, 79.5, etc I have changed list `L1` to reflect these boundaries. Ogives are most commonly displayed with percents (relative frequencies) on the *Y*-axis, although one can also use actual frequencies.

L2	L3	L4
0	0	------
0	6	
12	17	
14	8	
11	8	
1	1	
1	0	

L4 =...(L2)/40*100

We need to construct the cumulative frequency distribution(s). In the Statistics List editor, move the cursor to highlight the next empty list (`L4` or `list4` on the 89). On both calculators, we want to do list arithmetic, so on the 83/84 press [2nd][STAT] (`LIST`) on the 89 press [F3]. Now arrow to `OPS` and all calculators except the Nspire select choice `6:cumsum(`. If you are using the Nspire, find this command in the catalog. Enter the list name you want to sum (`L2` or `list2`), close the parentheses then type /40*100 to divide by the total number of observations, and then convert the decimal to a percentage. Press [ENTER] to complete the calculation.

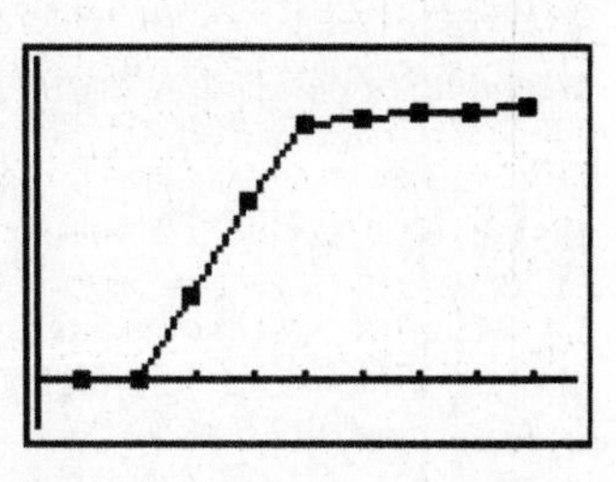

Define the plot using `L1` as `Xlist` and `L4` as `Ylist`. As with frequency polygons, on both calculators `Zoom Data` ([ZOOM][9] or [F5]) will display the graph. With [TRACE] activated, we see that 65% of the female pulse rates were at most 79.5 beats per minute.

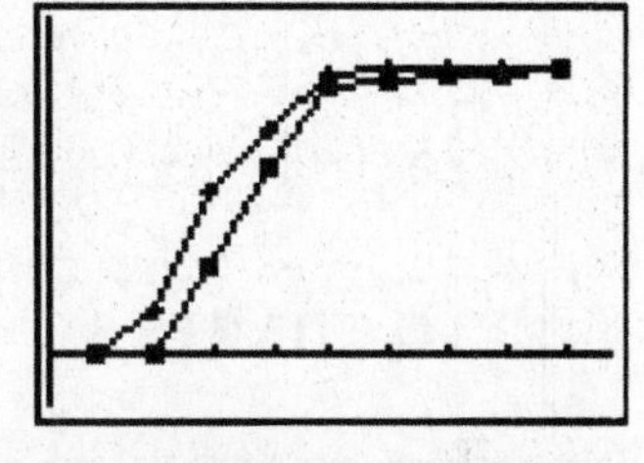

In the screen at right, I have repeated the process as described above using the male's data and defined a second plot as we did for the frequency polygons. Here we also see the left shift, indicating that these males, at least, tended to have lower heart rates than the females.

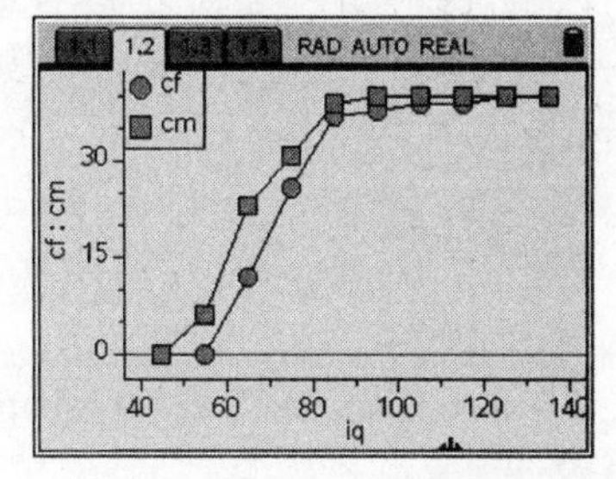

Nspire: In a like manner, follow the directions given for the frequency polygons above.

DOTPLOTS AND STEMPLOTS

We've already seen the TI-Nspire create a dotplot. Other TI calculators will not create dotplots, and none of them will create a stemplot. However, sorting the raw data can help in creating them "by hand."

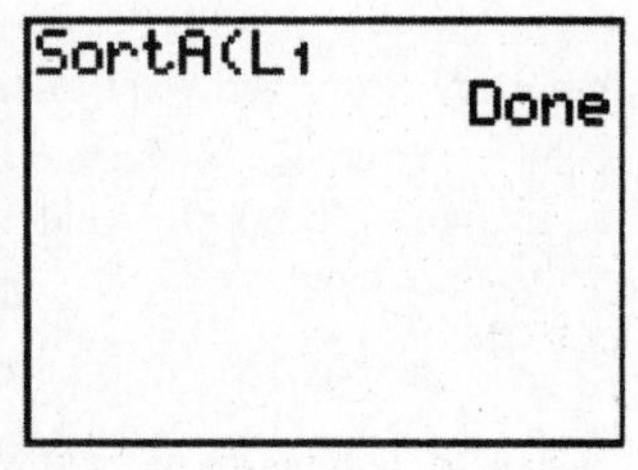

On the TI-84/84 series calculators, the Sort command is located on the initial [STAT] menu. There are two choices: `2:SortA(` and `3:SortD(` which sort the data in ascending (low to high) or descending order. In either case, select the appropriate choice by either pressing the desired number of the menu option or by arrowing to move the highlight and pressing [ENTER]. The command will be transferred to the home screen. Enter the name of the list to be sorted and finally press [ENTER] to execute the command. The calculator will tell you when it is finished by showing the message `Done`.

Return to the Statistics Editor and you can then page through the sorted list of data.

L1	L2	L3 1
60	0	0
60	0	6
60	12	17
64	14	8
64	11	8
64	1	1
64	1	0

L1(1)=60

On a TI-89 calculator, within the Statistics List Editor, select the List menu ([F3]) and arrow to Ops. Pressing ⊙ displays the menu. Press [ENTER] to select choice 1:Sort List. You enter the name of the list to be sorted and determine ascending or descending order. Pressing [ENTER] will execute the sort.

To sort a list with the Nspire, in the data spreadsheet, move the NavPad so the desired column is highlighted, then press the menu key followed by 1:Actions and 6:Sort.

PARETO CHARTS

Pareto charts are a special bar graph for categorical data, where the bars are sorted left to right from most often occurring to least often occurring. TI calculators cannot handle categorical data as such, but using a nominal (number) scale to represent the categories, we can create a Pareto Chart for the data in Exercise 13 of section 2-3. The chart at right reproduces the data from the text on relative frequency of college type.

Type	Frequency
Public 2-Year	36.8%
Public 4-Year	40.0%
Private 2-Year	1.2%
Private 4-Year	21.9%

Since there are four categories of complaint, enter the numbers 1 through 4 in L1, and the frequencies in L2. It is important that these be entered in descending order for the Pareto chart.

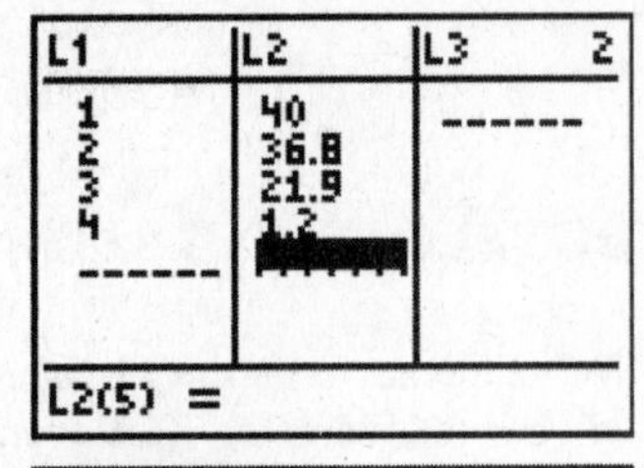

Set up the plot as a histogram with tabled data (use L1 as the Xlist and L2 as the frequency list). We want to see all the bars, and also specify that each bar has width 1. Ymin has been set to a low number so that using [TRACE] does not obscure the final plot. The same window settings will work on the TI-89, but be sure to set the histogram bucket size to 1.

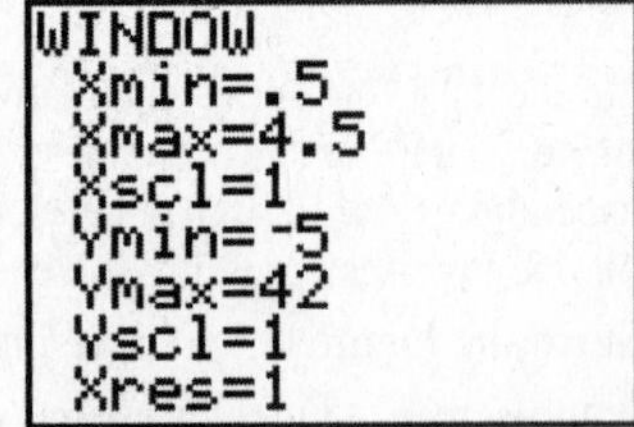

Pressing [GRAPH] followed by [TRACE] shows the following. Both public types have similar numbers; private 2-year colleges don't have many students.

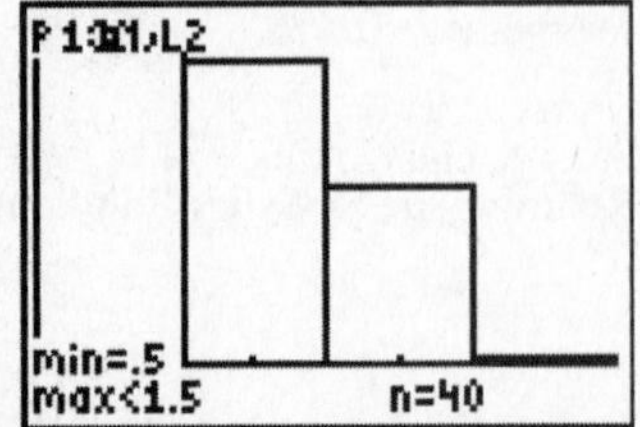

Nspire: Follow the directions for creating a histogram with tabulated data as given above; however, you will need to convert all frequencies to integers (in this case, multiply by 1000).

PIE CHARTS

Pie charts are not among the statistical plot types for TI Calculators, except for the Nspire. However, the calculator can aid in the making of these plots by calculating the degree measure of the center angle (portion of the entire 360° circle) for the pie wedge representing each category. We illustrate this using the data set of college types from Exercise 13. The instructions are analogous for both all calculators.

With the data on frequency of each complaint type already in L2, highlight list name L3. Type 360L2[÷]sum(L2 as in the bottom of the same screen with "sum(" being pasted in by pressing [2nd] [STAT][▸] [▸] [5] (It is the fifth choice on the LIST<Math> menu). On the TI-89, access the List menu with [F3].

In the screen at right, we see that public college account for more then 75% of all undergraduate students. The central angle for private 2-year colleges would be only 4.3 °.

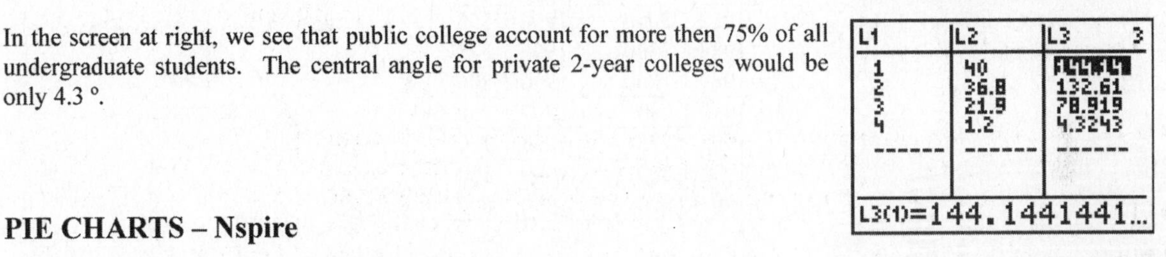

PIE CHARTS – Nspire

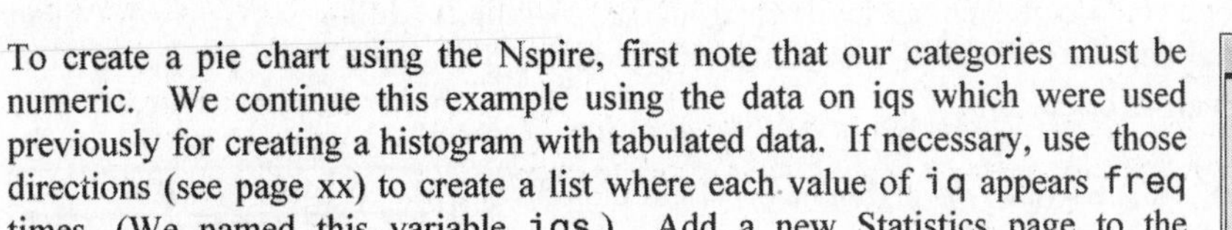

To create a pie chart using the Nspire, first note that our categories must be numeric. We continue this example using the data on iqs which were used previously for creating a histogram with tabulated data. If necessary, use those directions (see page xx) to create a list where each value of iq appears freq times. (We named this variable iqs.) Add a new Statistics page to the worksheet by pressing (⌂)(5). Move the cursor to the lower part of the window so you can click to add a new variable. Add variable iqs. This creates a plot that looks like a bar chart (but is really a dot plot – there are many repetitions of each value).

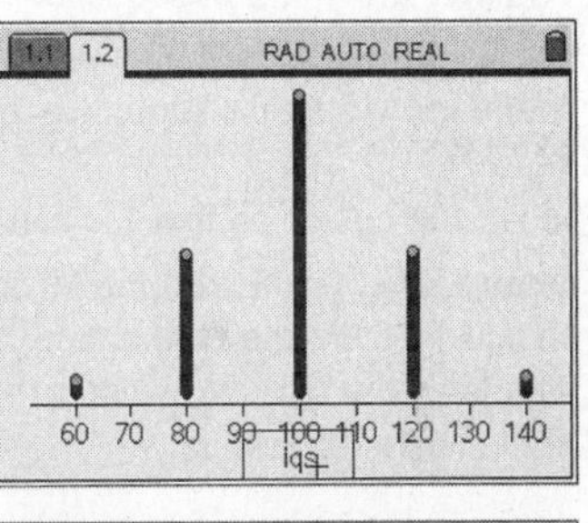

We transform this graph into a pie chart by pressing (menu), (2) for Plot Properties, then (8) Force Categorical X. This changes the plot into a bar chart as at right. (Bars are sorted in ascending order by the first character of the value).

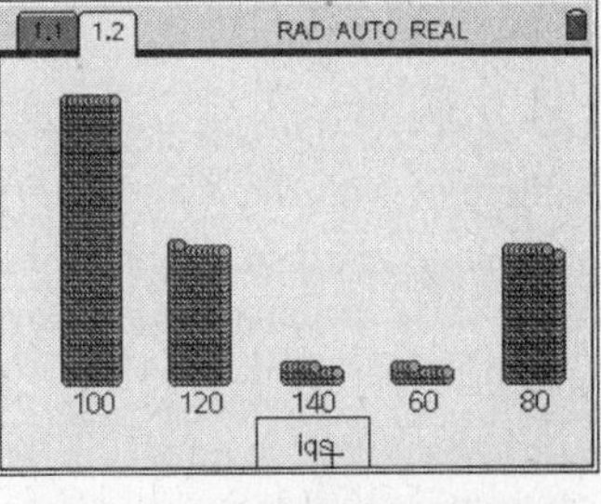

Finally, change this plot to a pie chart by pressing (menu), (1) for Plot Type, and (9) for Pie Chart.

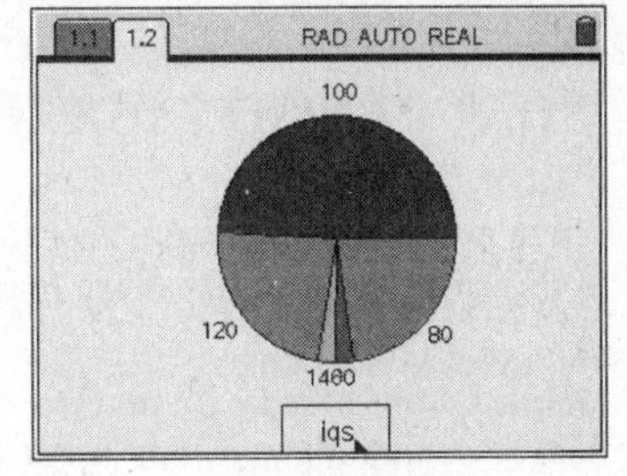

SCATTERPLOTS

Scatterplots will be discussed in Chapter 10 on Linear Regression.

TIME-SERIES GRAPHS

Time-series graphs show how a single variable has changed with the passage of time, as opposed to histograms or frequency polygons which show a "snapshot" of the distribution of the variable at one specific time. These are constructed as connected plots, just as frequency polygons are. The time index is always displayed on the *X*-axis, and the variable of interest on the *Y*-axis. The table below gives the number of drive-in theater screens in the country for years 1872 through 2000.

Year	Screens	Year	Screens
1987	2050	1994	850
1988	1500	1995	825
1989	1000	1996	800
1990	900	1997	750
1991	875	1998	710
1992	860	1999	700
1993	800	2000	600

To enter the years in sequence, one could type them all in, or enter the last two digits (calling 2000 year 100), but the easiest way to enter sequential data uses the seq(function which is choice 5 on the List Ops menu. The command works the same way on all TI calculators.

Move the cursor so that the name of list L1 (list1) is highlighted. Go to the List menu (2nd STAT on the 83/84 or F3 on the 89), arrow to Ops, then select menu choice 5. Type the command: X,X,1987,2000). Pressing ENTER populates the list. Now enter the number of drive-in screens in L2. The screen at right shows the results.

L1	L2	L3 2
1987	2050	------
1988	1500	
1989	1000	
1990	900	
1991	875	
1992	860	
1993	800	

L2(1)=2050

My statistics plot was defined just as we did for frequency polygons: a connected scatterplot with the Xlist being L1, the Ylist L2. I also selected to use the single pixel point mark so as not to obscure the graph. Pressing F5 on the 89 or ZOOM 9 on an 83/84 displays the plot at right.

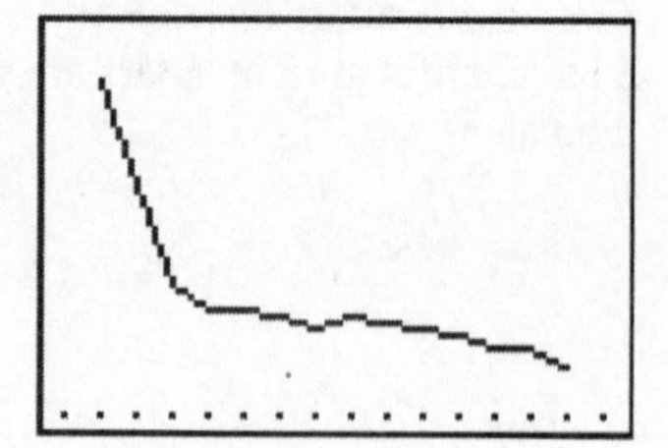

WHAT CAN GO WRONG?

There are several common error messages which might show and "foul-ups" which can occur when graphing data. In this section, I describe those most frequently encountered.

Help! I can't see the picture!

Seeing something like this (or a blank screen) is an indication of a windowing problem. This is usually caused by pressing GRAPH using an old setting. Try pressing ZOOM 9 to display the graph with the current data. This error can also be due to having failed to turn the plot "On."

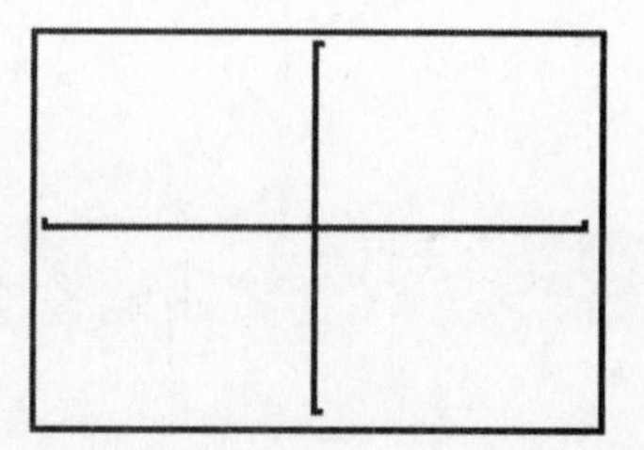

What's that weird line (or curve)?

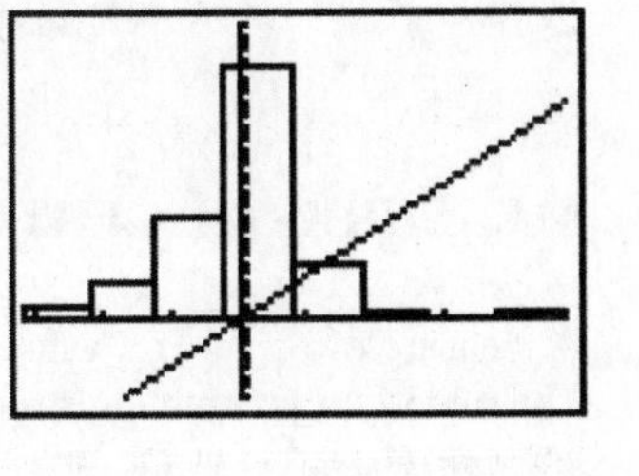

There was a function entered on the [Y=] screen. The calculator graphs everything it possibly can at once. To eliminate the line, press [Y=]. For each function on the screen, move the cursor to the function and press [CLEAR] to erase it. Then redraw the desired graph by pressing [GRAPH].

What's a Dim Mismatch?

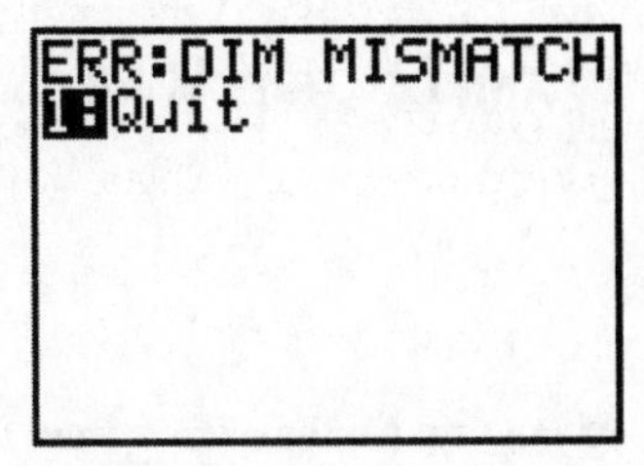

This common error results from having two lists of unequal length. Here, it pertains either to a histogram with frequencies specified or a time plot. Press [ENTER] to clear the message, then return to the statistics editor and fix the problem.

What's an Invalid Dim?

ERR:INVALID DIM
1:Quit

This problem is generally caused by reference to an empty list. Check the statistics editor for the lists you intended to use, then go back to the plot definition screen and correct them. On the TI-89 this error message is Dimension.

What's Err:Stat?

ERR:STAT
1:Quit

This error usually happens when you press [ZOOM][9]. It is caused by having two plots turned on at the same time that have incompatible data values (one list has much larger numbers than another). Turn off the extra plot.

3 Statistics for Describing, Exploring, and Comparing Data

MEASURES OF CENTER

A measure of center is a value at the center or middle of a data set. They describe a "typical" value for a data set. There are four common measures of center: the mean, median, mode, and midrange. All of these are at one time or another referred to as the "average." You should be specific which measure you are using.

EXAMPLE Men's Word Counts: Listed below are five measurements of the number of words spoken daily by men.

27,531 15,684 5,638 27,997 25,433

Mean and Median from Raw Data – TI-83/84

```
1-Var Stats L1
```

Enter the data in L1 (do not enter the commas). Press STAT, then press ▶ to move to the CALC submenu, then press 1 for 1:1-Var Stats, then press 2nd 1 to name the list L1 as the list for which you want the statistics.

```
1-Var Stats
 x̄=20456.6
 Σx=102283
 Σx²=2466400359
 Sx=9670.030264
 σx=8649.138006
↓n=5
```

Press ENTER for the first portion of the output. You know this is only the first portion, because there is the ↓ on the lower left corner, which is an indicator of more output available. From this first screen, we find the mean (arithmetic average) of the word counts is $\bar{x} = \frac{\Sigma x}{n} = 20,456.6$ (the usual "rule of thumb" is to report one more decimal place than was in the original data, unless your instructor says otherwise).

```
1-Var Stats
↑n=5
 minX=5638
 Q1=10661
 Med=25433
 Q3=27764
 maxX=27997
```

To reveal the second output screen, hold down the ▼ key. Here we find the median = Med = 25,433. Many of the other statistics displayed on the two screens will be discussed later.

Mean and Median from Raw Data – TI-89

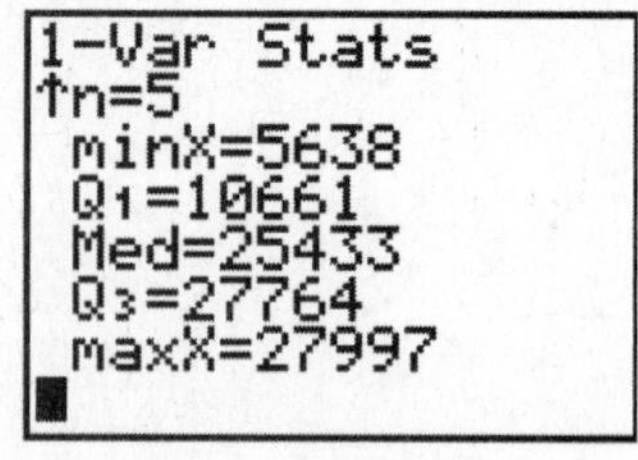

Enter the data in List1. Press F4 (Calc), then press ENTER to select 1:1-Var Stats. Press 2nd − (VAR-LINK) to and find the list name List1 as the list for which you want the statistics. Press ENTER to paste the list name into the input screen. Freq should be set to 1, as each value in the list occurred once. Your input screen should look like the one at right.

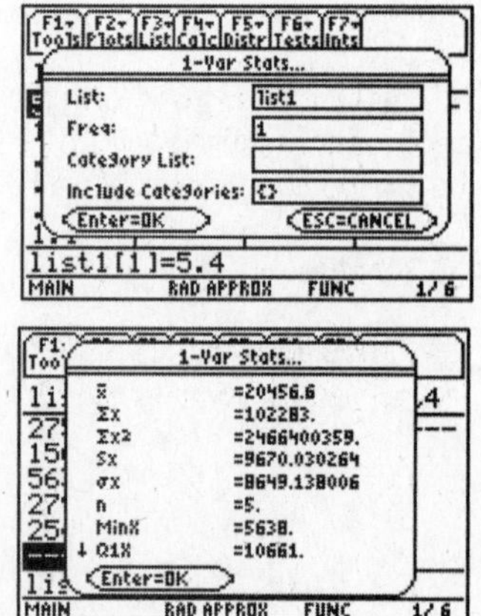

Press ENTER for the first portion of the output. You know this is only the first portion, because there is the ↓ on the lower left corner, which is an indicator of more output available. From this first screen, we find the mean (arithmetic average) of the lead readings is $\bar{x} = \frac{\sum x}{n} = 20{,}456.6$ (good statistical practice is to report one more decimal place than was in the original data, unless your instructor says otherwise).

To reveal the second output screen), hold down the ▼ key. Here we find the median = Med = 25,433. Many of the other statistics displayed on the two screens will be discussed later.

Mean and Median from Raw Data – TI-Nspire

Enter the data in a column in the spreadsheet. Press (menu), then 4:Statistics, 1:Stat Calculations, 1:One-Variable Statistics. We have a single list for statistics, so on the intermediate screen, press (enter). My data were in column a. Each has frequency 1. Note that the first result column (the statistic names) will be column b. If you want them somewhere else, type a new column name. Press (enter) to perform the calculations.

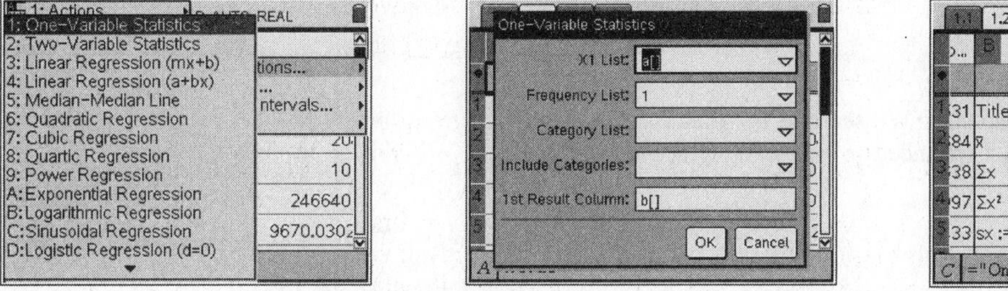

We really can't see the statistics. We'll resize column C. Press (menu), `1:Actions`, `2:Resize`, `2:Maximize Column Width`. Just as with the other calculators, arrow down to find the median.

Mode

TI calculators do not automatically calculate the mode of a data set. Ordering the data and counting the frequency of each value, as discussed in the section on dotplots, will aid in finding the mode (if one exists). The modal class can be found by looking for the class with the highest frequency in a frequency table. (See page 22 of this manual.)

Midrange

EXAMPLE Men's Word Counts: Repeated below are the five measurements of the number of words spoken daily by men. Find the midrange.

27,531	15,684	5,638	27,997	25,433

Two of the values on the second output screen are `Min` and `Max`. These are the lowest and highest values in the data set. We can use the output displayed above to calculate midrange $= \dfrac{\text{Max} + \text{Min}}{2} = \dfrac{27997+5638}{2} =$ 16,817.5.

Mean from a Frequency Distribution

Word Count	Frequency	**Midpoint**
0–9,999	46	4999.5
10,000-19,999	90	14999.5
20,000-29,999	40	24999.5
30,000-39,999	7	34999.5
40,000-49,999	3	44999.5

When data is presented as a frequency distribution, the actual data values are lost, but one can approximate the mean and other summary statistics. Table 3-1 gives the frequency distribution of the number of words spoken by the 186 men in the sample. Enter the midpoints in `L1` (`list1`) and the frequencies in `L2` (`list2`).

On a TI-83/84, press [STAT] then [▶] to CALC and press [ENTER] to select `1:1-Var Stats`. We need to tell the calculator to use both lists, but we still want statistics for only one variable (the number of words). To do this, specify the list names `L1,L2` ([2nd][1][,][2nd][2]) as at right.

```
1-Var Stats L1,L
2█
```

Press [ENTER] to see the results. . We see that the mean value calculated from the frequency table is 15,913.5 (rounded to one more place than was in the original data).

```
1-Var Stats
 x̄=15913.47849
 Σx=2959907
 Σx²=6.1047E10
 Sx=8681.955084
 σx=8658.585042
↓n=186
```

On a TI-89 calculator, to use frequency tables in the calculation, use `1-Var Stats` from the `Calc` menu as before, but specify the second frequency list in the box labeled `Freq`. See the example at right.

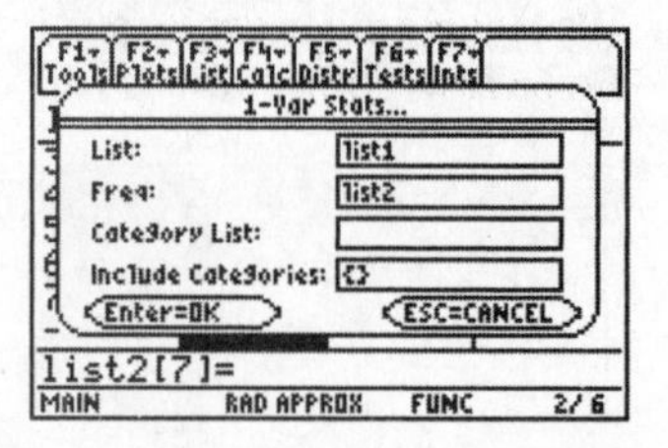

With an Nspire, be sure the variable columns have names at the top, and select the proper variable names using the down arrow key.

Weighted Mean

EXAMPLE: Find the mean of three tests scores (85, 90, 75) if the first test counts 20%, the second counts 30%, and the third test counts for 50% of the final grade.

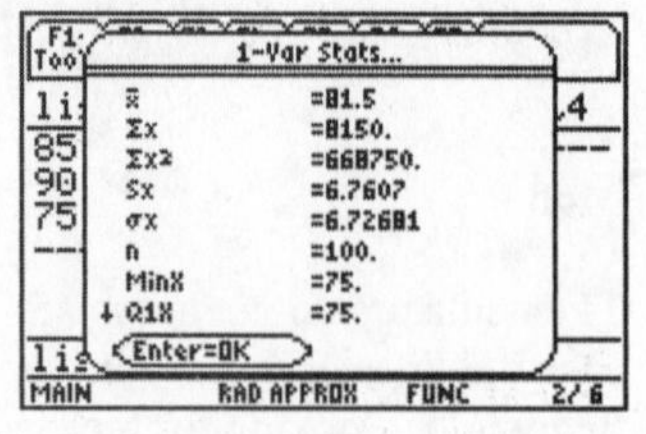

Put the scores in `L1` (`list1`) and the weights in `L2` (`list2`), and compute as if this were a frequency distribution as described above. This student has a grade of 81.5, so a B (at least in my grading system).

MEASURES OF VARIATION

A measure of variation indicates in some manner how spread out or close together a set of data might be. Data sets with large variability indicate less consistency in the values.

EXAMPLE : As of this writing, the number of satellites for military and intelligence purposes owned by India, Japan, and Russia are 1, 3, and 14. Find the range, standard deviation and variance of the number of satellites.

Standard Deviation

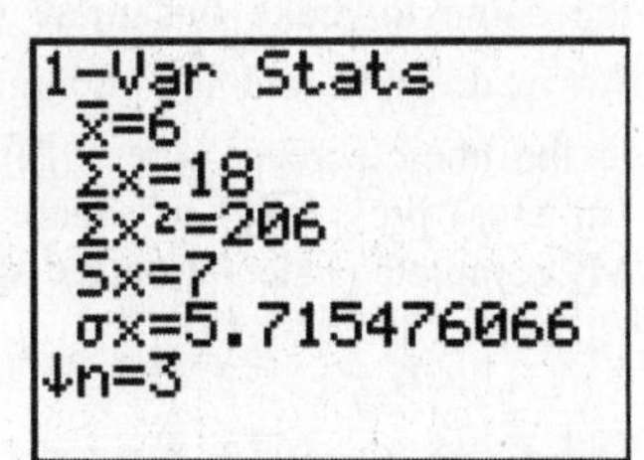

With the data in L1, press [STAT], arrow to CALC, press [ENTER] to select 1:1-VarStats and enter the list name ([2nd][1]). The standard deviation is Sx = 7 satellites. The σx value of 5.7 would be used if we had data for the entire population (the number of satellites owned by all countries, but we don't have this —just a sample.) On a TI-89, use 1-Var Stats as described above.

Variance

The variance is not among the statistics given by the 1-Var Stats display, but it is easy to calculate since it is the square of the standard deviation. In this case, Variance = $(Sx)^2 = 7^2 = 49$ satellites^2. Note that the units here (satellites^2) does not make sense – this is why people normally do not use this statistic, but its square root.

Range

The range is given by maxX - minX = 14 – 1 = 13 satellites.

Using Statistics Variables

Sometimes it is useful to be able to recall statistics *that have already been calculated.* One example of this is in computing the variance of a distribution. If there are many digits in the standard deviation, retyping them could lead to typos and erroneous results. We also want to keep as many decimal places as possible in intermediate calculations and do any rounding at the final result. It is important to know that you can only paste a statistics variable after you have performed 1-Var Stats on your *current* data set, otherwise the statistics stored will be from some past calculation.

TI-83/84 Procedure

Press [VARS] for the VARS menu screen, then press [5] for the Statistics sub menu screen at right.

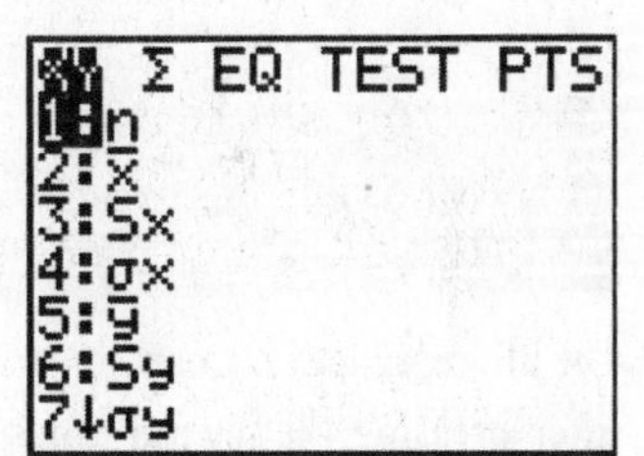

Press [3] and Sx is pasted onto your home screen. Now, press [ENTER] to see that Sx is 7 as in the top of my screen. Press [x^2] and then [ENTER] to again see that the variance = $s^2 = 49$.

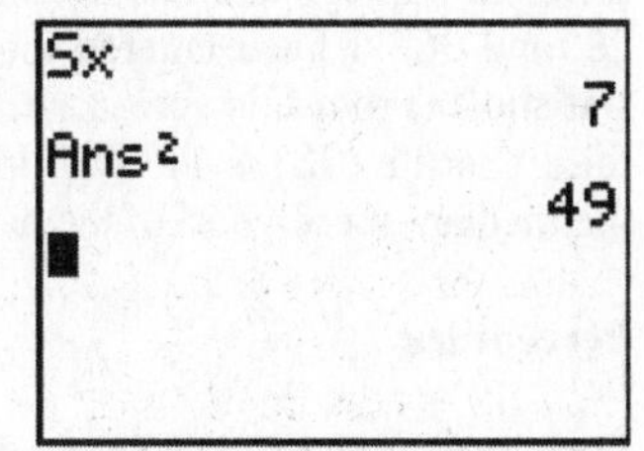

TI-89 Procedure

From the Home screen, press [2nd][-] ([VAR-LINK]). The MAIN folder is normally expanded. Arrow down to highlight the folder name, and press ◀ to collapse that list. You should see a screen like the one at right.

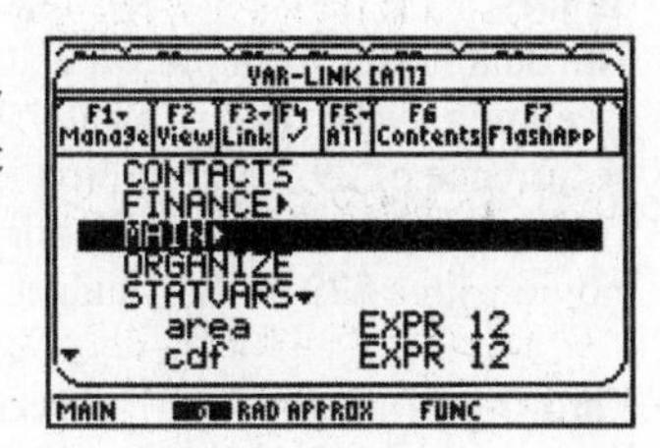

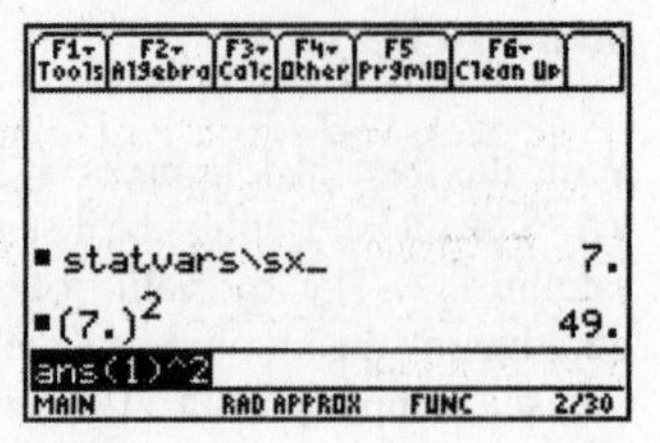

Arrow down to find the statistics variable you want. You can type the first letter fo the name to make this more efficient. The sample standard deviation is sx_. Arrow down to find and highlight it, then press [ENTER] to transfer the variable name to the home screen. Press [ENTER] again to display the value. Now (to find the variance) press [^][2] to square the standard deviation and compute the variance. My completed calculation is displayed at right.

Standard Deviation from a Frequency Distribution

On page 27 we estimated the number of words spoken by a man using a frequency distribution. Our output screen is duplicated at right. From this we estimate the standard deviation of the number of words is 8682.0. Bear in mind that just as the mean in this instance is an estimate of the true value that would be obtained from the entire data set, so is the standard deviation an estimate.

MEASURES OF RELATIVE STANDING

Quartiles and Percentiles

Table 3-4 lists the sorted budget amounts (in millions of dollars) for a simple random sample of 35 movies. These are reproduced here for convenience.

4.5	5	6.5	7	20	20	29
30	35	40	40	41	50	52
60	65	68	68	70	70	70
72	74	75	80	100	113	116
120	125	132	150	160	200	225

```
1-Var Stats
↑n=35
 minX=4.5
 Q1=35
 Med=68
 Q3=113
 maxX=225
```

On all calculators, quartiles are displayed on the 1-Var Stats output screen after pressing [▼] several times to scroll to the bottom. Your text illustrates the different implementations for finding quartiles using data values 1, 3, 6, 10, 15, 21, 28, and 36. I have entered those values in list1 and computed the summary statistics. From this screen we find the first quartile (Q1) is 35 (million) and the third quartile (Q3) is 113 (million). TI calculators all use the algorithm of finding the median of each half of the data set after it is split by the median.

Percentiles

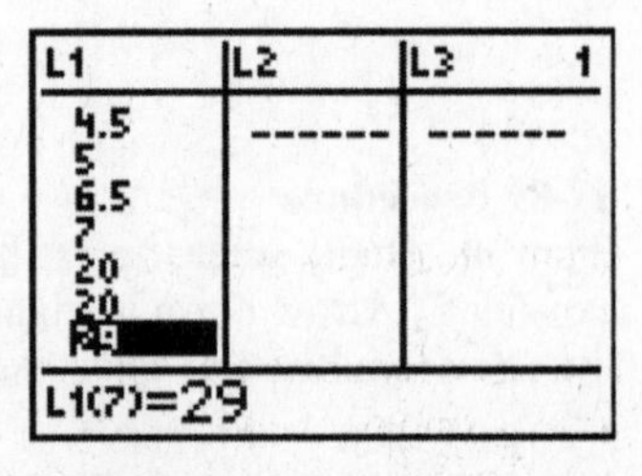

Find the percentile corresponding to a movie with a budget of 29 million dollars.
The data in this example were already sorted; if not, we would first sort the data in ascending order (see Chapter 1). Now scroll down the list to find the first occurrence of 29. \$29 million is the 7th entry, so there are 6 movies that cost less. There are 35 total movies in our list, so we calculate (6/35)*100 = 17.1. Thus, an movie with a \$29 million budget is the 17th percentile of this distribution.

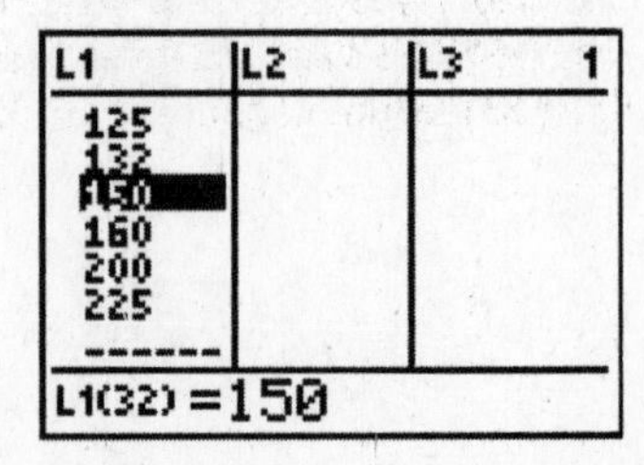

Find the value of the 90th percentile or P_{90}.
We calculate (90/100)*35 = 31.5. Since this is not a whole number, we round up to the nearest whole number which is 32. Thus the 90th percentile is the 32nd value in our sorted list. We find this value to be a budget of \$150 million.

Find the third quartile Q_3.
If we look back at the calculator's Q_3 statistic displayed in the `1-Var Stats` calculations, we see that the value is Q3 is 39.5. Alternatively, since $Q_3 = P_{75}$, we could proceed as in the previous example. We find (75/100)*35 = 26.25. We make this 27 and find that Q_3 is $113 million (which agrees with what `1-Var Stats` found).

L1	L2	L3 1
113		
116		
120		
125		
132		
150		
160		

L1(27) =113

BOXPLOTS AND THE FIVE-NUMBER SUMMARY

Boxplots are another way to picture a distribution. There are two boxplots implemented on TI calculators: the skeletal or regular plot which uses only the five-number summary (min, Q_1, med, Q_3, and max) to graph the "quarters" of the distribution and the modified boxplot which identifies outliers. They are the fourth and fifth plot types on a TI-83 or -84 (icons for the regular boxplot and for the modified boxplot). On a TI-89 these are choices `3:Box Plot` and `5:Mod Box Plot` on the Plot definition screen. The Nspire defaults to the modified boxplot. Select this from the Plot Types menu option on any single variable plot. Boxplots are usually very good at finding "typos" in a list of data, and they are extremely useful in comparing two distributions.

EXAMPLE Movie Budgets with a typo
I have used the data set on movie budgets, but replaced one occurrence of a 70 with 7070 (a distracted typo). We notice first of all the impact of this typo – the average cost is not 274.1 million (more than the most expensive of our films), and the standard deviation is 1184! Obviously, thinking about movie budgets, these are not reasonable values – go back and double check your data entry!

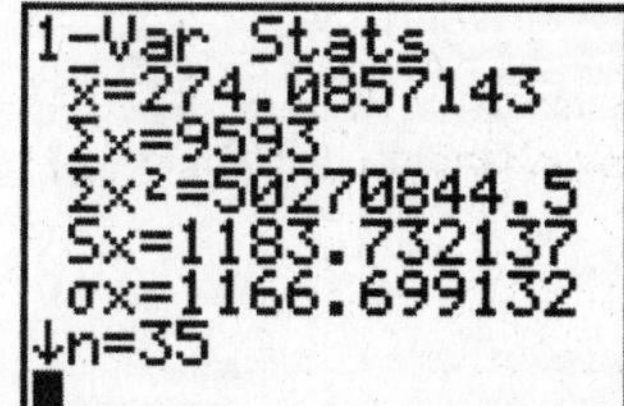

Boxplot (or Box and Whisker Diagram)

I have corrected the typo for this example.

Press [2nd] [Y=] ([F2] on an 89) to get the `Stat Plots` menu and then [1] to get the set-up for `Plot1`. Set up `Plot1` as shown in my screen. We choose the fifth plot type.

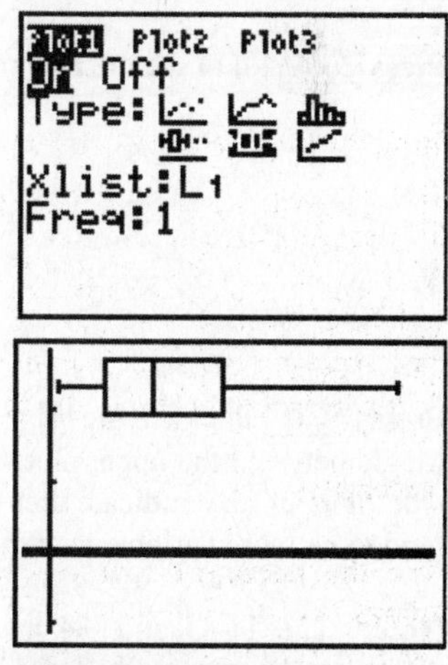

Press [ZOOM] [9] for the boxplot. Press [TRACE] then use the [◄] and [►] keys to display the **five-number summary**.

Identifying Outliers – The Modified Box Plot

The plot obtained above has a long right-hand tail. It may be that there are outliers present (movie budgets that don't fit the pattern of the rest of the data). TI calculators can create the modified box plot which labels any observation more than 1.5*IQR away from the quartiles as an outlier. In the screen at right, I have changed the plot definition to make a modified box plot.

Pressing ZOOM 9 (F5 - ZoomData on an 89) and then TRACE shows that there are no outliers in this particular data set. The long right tail is just a long tail.

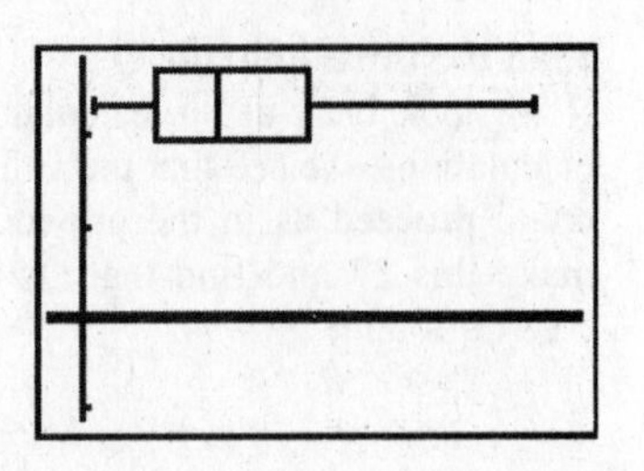

Comparing Data Sets with Side-by-Side Boxplots

We have said before that normally there should be only one statistics plot "on" at a time. Boxplots for comparing two distributions are a good example of a reason to have more than one plot "on" at once.

EXAMPLE Pulse Rates of Men and Women: We compare the pulse rates of 40 females and 40 males as seen in Data Set 1 in Appendix B. We have stored the females' rates in L1 and the males' rates in L2.

Press 2nd Y= to get the Stat Plots menu and then define Plot1 as at right.

Press 2nd Y= to get the Stat Plots menu and then define Plot2 as at right.

Press ZOOM 9 to see the two boxplots on a single screen.

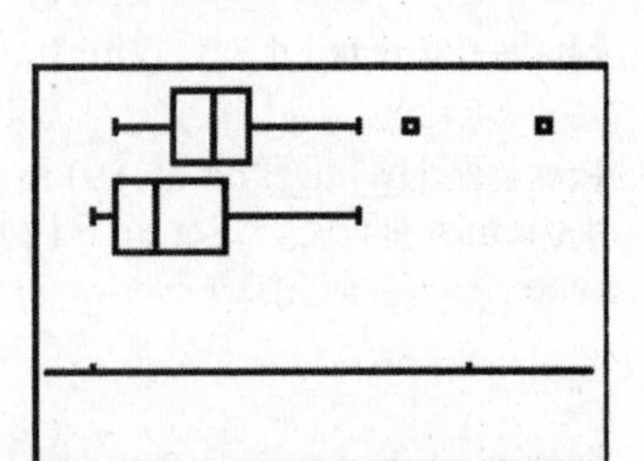

Press TRACE then use the ◄ and ► keys to display the five-number summary. You can use the ▲ and ▼ keys to toggle between the upper plot which is that of the females and the lower plot which is for the males. These side-by-side plots clearly indicate that males, in general, have pulse rates than females and that the pulse rates of females tend to be more variable than those of males.

Note: The TI-83 and -84 have three StatPlots. They can be used to plot box and whisker plots for three different data sets on the same graph. The TI-89 can display as many as six boxplots at once. Be sure to turn "off" the extra plots when you're done with them!

Side-by-side Boxplots on the Nspire

To create side-by-side boxplots on this calculator, we need to use its window splitting capability, then (possibly) change the axis scales so they are the same. Add a statistics plot page, then click to add the variable (column) named women. You first have a dotplot. Now, press menu, then 1:Plot Type and change to a boxplot. Now, we'll split the window. Press ctrl home for the Tools menu. Select option 5:Page Layout, 2:Select Layout, then 3:Layout3.

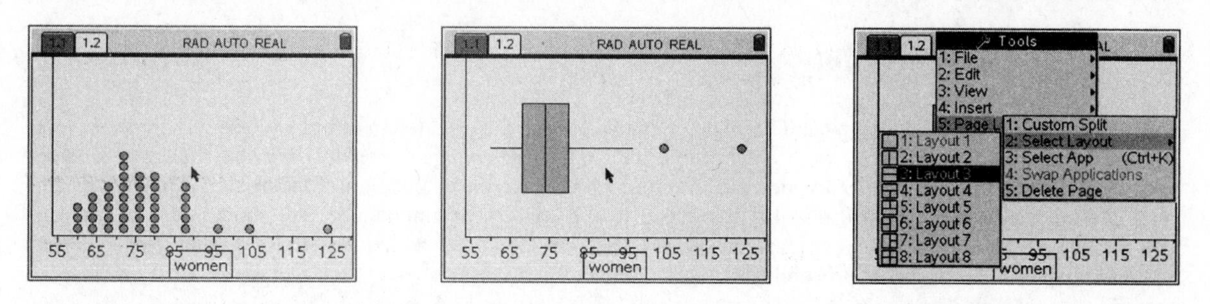

The window splits horizontally, with the women's graph in the top half. Now, move the arrow cursor into the bottom window, click the grabber key to make that the active window. Press ⌂ and add a Data and Statistics app in the bottom window. Move the cursor and click to add men to that window, then press menu, 1:Plot Type and change this to a boxplot. We have the two plots, but the scales don't match. We'll change the men's scaling to match the women's. Press menu, 4:Window/Zoom, 1:Window Settings. We've changed Xmin to 55 and Xmax to 130. The resulting graphs are shown below.

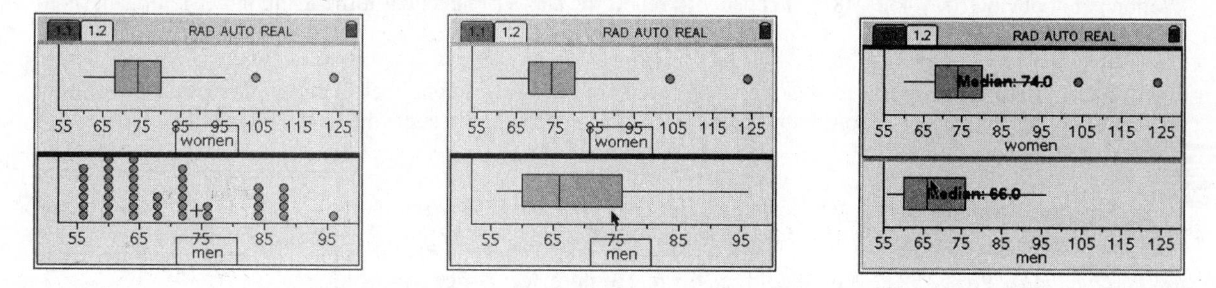

WHAT CAN GO WRONG?

Primarily, errors here are due to data entry mistakes. Always double check what you have entered. If a boxplot fails to display, the plot more than likely was not turned on. Always be sure to turn off any extra plots after copying them to paper. If not, you probably will receive either the `Invalid Dim` (referring an empty list) or `Stat` (incompatible window ranges for two plots) error messages. These were discussed in the previous chapter.

4 Probability

In this chapter you will learn how your TI calculator can be used in the calculation of probabilities. There are several built-in functions which make this work much simpler, including the *change to fraction function,* the *raise to power key,* and the *Math Probability menu* which includes options to make the calculation of factorials, combinations and permutations very easy. You will also be introduced to the concept of simulation. This process can be used to approximate probabilities. Your exposure to the process may have the added benefit of giving you a deeper understanding of the concepts of probability.

TWO USEFUL FUNCTIONS

Change to Fraction Function

This function is useful when one wants to display an answer in fraction, rather than decimal form. However, when a fraction is not obvious (say 432/7482), it is better to report the answer in decimal form, using three significant (non-zero) digits.

EXAMPLE Birth Genders: In reality, more boys are born than girls. In one typical group, there are 205 newborn babies, 105 of whom are boys. If one baby is randomly selected from the group, what is the probability that the baby is *not* a boy?

```
100/205
               .487804878
Ans▸Frac
                    20/41
```

Using the Law of Complements, we decide the answer is 100/205=0.488. See the first line of the screen at right. Now, press [MATH] [1] (this chooses the *change to fraction function).* Press [ENTER]. You should see the rest of the screen. The result is the probability written as a reduced fraction $\frac{20}{41}$.

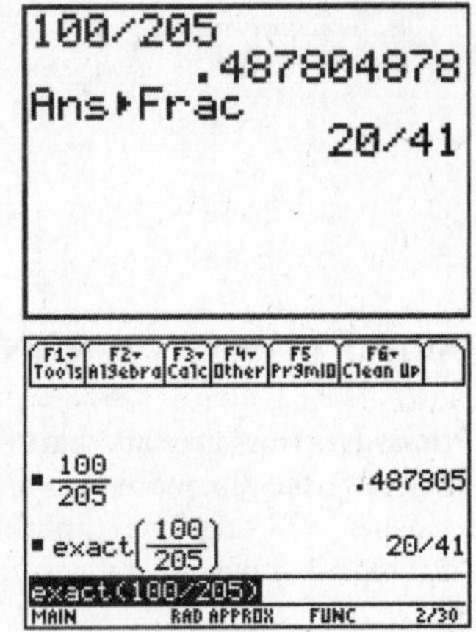

On a TI-89, we recommended in Chapter 1 that the mode for answers be set to `Approximate` rather than the calculator's default of `Exact`. To obtain exact answers with the calculator in `Approximate` mode, Press [2nd][5] (Math), then press ▸ to expand the `Number` menu. Press [ENTER] to select `exact(`. Now type in the division, being sure to close parentheses before pressing [ENTER] to complete the calculation. The TI-89 equivalent of the TI-83 calculation above is at right. If a TI-Nspire is set to auto in document or system settings, it will display the reduced fraction automatically.

The Raise-to-the-Power Key
The Probability of "at Least One".

EXAMPLE Gender of Children: Find the probability of a couple having at least one girl among three children. We will assume for the purpose of this example that boys and girls are equally likely. We make use of the complements rule, since all boys is the opposite (complement) of at least one girl.

```
.5^3
                .125
1-Ans
                .875
Ans▸Frac
                 7/8
■
```

Find the probability of the complement. P(boy and boy and boy) = 0.5*0.5*0.5 = 0.5^3. Type 0.5 [^] [3]. Then press [ENTER] for 0.125.

Note: Keep in mind the answer you have now (the probability of all boys) is the complement of the answer you seek. Press [1] [−] [2nd] [(-)]. This takes your answer from above and subtracts it from 1. Press [ENTER] for 0.875. This could be changed to the fraction 7/8 as shown.

PROBABILITIES THROUGH SIMULATION

Finding probabilities of events can sometimes be difficult. We can often gain knowledge and insight into the problem by developing a simulation of it. The techniques used in the examples in this section build upon one another. Thus each example assumes you are familiar with the ones preceding it.

EXAMPLE Gender Selection: When testing techniques of gender selection, medical researchers need to know probability values of different outcomes, such as the probability of getting at least 60 girls among 100 children. Assuming that male and female births are equally likely, describe a simulation that results in the gender of 100 newborn babies. We will perform this simulation by generating 100 random 0's and 1's. Each will be equally likely.

TI-83/84 Procedure:

We begin by setting the "seed" for our random number generator as 136. (The reason for this step is to set your calculator, so it will generate the same random numbers as generated by the calculator used in this manual. This step is **not** necessary when performing your own simulations.) Press 136 [STO▶] [MATH], arrow to PRB, press [1] to select `rand`, then [ENTER] as shown at the top of my screen.

```
136→rand
                     136
randInt(0,1,100)
→L1
{0 0 1 1 0 1 1 ...
sum(L1)
                      52
```

Type [MATH] [◀] [5] to get the `randInt` prompt on the main screen. Then type 0,1,100) [STO▶] [2nd] [1] [ENTER] to have the TI-83/84 generate one hundred 0's and 1's and store them in L1. We can see the results which begin with {0 0 1 1 0 1 1, ...} These can be translated as {B B G G B G G...}if we let the 1's represent girls and the 0's represent boys.

In the last line of the screen above, we summed the elements in list L1 to find out how many 1's (girls) we had in this simulation of 100 births. Press [2nd] [STAT] for the LIST menu, then [▶] to Math, select menu option `5:sum(`, then [2nd] [1] [ENTER] to sum the elements of list L1. We see that this time we had 52 girls.

Note: We did not have 60 or more girls in this simulation. In order to get an idea of the probability of at least 60 girls in 100 births, you would have to perform the simulation repeatedly, keeping track of how often the event of 60 or more girls occurred.

TI-89 Procedure:

With the calculator in the Statistics/List Editor application, press [F4] (`Calculate`) and either arrow to `4:Probability` or press [4]. To set the seed for the random number generator, the menu option is `A:RandSeed`. You can arrow down to find it, but it is easier to press ⊖ once.

```
1:1-Var Stats...
3:nCr(
4:!
5:randInt(
6:.randNorm(
7:randBin(
8:randSamp(
9:rand(
A:RandSeed...
```

Type in the desired seed, and press [ENTER].

Now, move the cursor to highlight the name of an empty list, press [F4] (Calculate). Select menu option 4:Probability as before, but now select option `5:randInt(`. Now type 0,1,100) - remember to close the parentheses - and press [ENTER] to populate the list.

```
list1  list2  list3  list4
0.
0.
1.
1.
0.
1.
list1[1]=0.
MAIN   RAD APPROX   FUNC   1/6
```

To sum the list, move the cursor to the top of an empty list. Press [F3] (`List`), arrow to menu option 3:Math and press ▶ to expand the menu. Select menu option `5:sum(`. Now we need to tell the calculator which list to sum. Press [F3] again, then [ENTER] to select `1:Names`. Find the name of the appropriate list, move the highlight to it, close the parentheses, and then press [ENTER] to complete the calculation.

TI-Nspire

The procedure is similar on the Nspire. Move the cursor to the formula area of a blank column in the spreadsheet, and type the = sign. Press and locate the `Probability, Random` set of functions. You may need to press the right arrow key to expand the catalog section. Locate `Random Integer` and press to transfer the command shell. Fill in the parameters of low, high, and number of trials; press to generate the list.

EXAMPLE Same Birthday: One classic exercise in probability is the *birthday problem* in which we find the probability that in a class of 25 students at least 2 students have the same birthday. Ignoring leap years, describe a simulation of the experiment that yields birthdays of 25 students in a class.

We will perform this simulation by generating 25 random integers between 1 and 365 (representing the 365 possible birthdays in a non-leap year.)

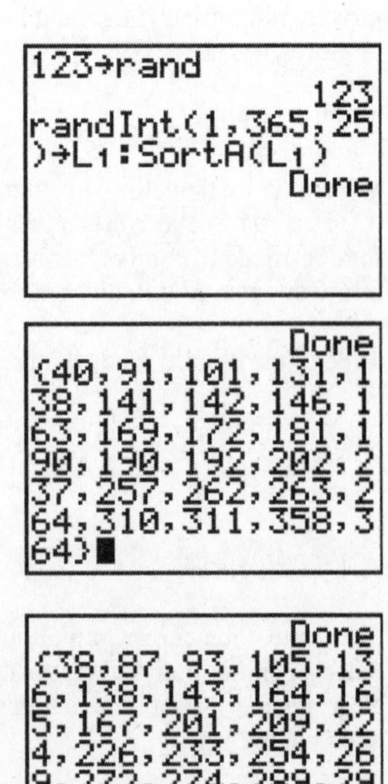

Again we set the seed so as to obtain the same outputs as this manual. Press 123 [STO▶] [MATH] [◄] [1] [ENTER].

Next generate 25 random integers between 1 and 365, store them in L1and sort L1. Do this by typing [MATH] [◄] [5] to get the `randInt(` prompt then typing 1, 365, 25 [STO▶] [2nd] [1] [ALPHA] [.] [STAT] [2] [2nd] [1] [ENTER] to see the Done message as in my screen. Note the [ALPHA] [.] sequence yielded a colon on the screen. The colon can be used to tie several statements together in one command.

We wish to see the results of our simulation, so recall L1 to the screen with [2nd] [STO▶] [2nd] [1] and then press [ENTER] for screen the screen at right. We can see that two students in this simulated class had the same birthday on the 190th day of the year.

To repeat this simulation without retyping all the commands, simply press [2nd] [ENTER] [ENTER]. This sequence yields the ENTRY command which repeats the last command line. Now press [2nd] [STO▶] [2nd] [1] to recall the new list L1 as in the screen at right. Press [ENTER] to check the next set of results. It shows no birthday matches in this simulated group of 25 students.

To do this on a TI-89, follow the steps outlined above using the TI-89 procedure previously described. To sort the list, place the cursor so it highlights the list name, press [F3] (`List`), arrow to option `2:Ops`, then press [▶] to expand the menu. Press [ENTER] to select `1:Sort List`. The dialog box should look like the one at right (assuming your random numbers were in `list1`.) Page through the list looking for any duplicates.

Thus far we have seen a 50% chance of a birthday match in a group of 25. Naturally, the simulation should be repeated many more times to get a better estimate of the actual probability of this event.

EXAMPLE Simulating Dice: Describe a procedure for simulating the rolling of a pair of six-sided dice.

This time we seed the random number generator with 4321. (Again, this step is only necessary if you are trying to duplicate the results presented in this guide.)

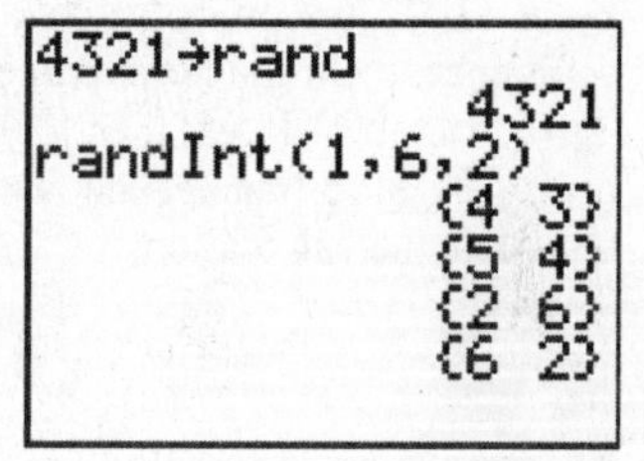

To generate two random integers between 1 and 6 type [MATH] [◀] [5] to get the `randInt(` prompt then type 1, 6,2 and close the parentheses. Press [ENTER]. If you used the seed from step 1, you should see the result (4,3) as at right. Continue to press [ENTER] to generate new rolls as shown in the bottom of the screen. TI-89 calculators cannot perform this type of random integer generation on the home screen, but must use statistics lists with the `randInt(` function. See the examples in Chapter 1.

To quickly generate 100 sets of two rolls and calculate their sums, you could proceed as follows. This procedure is analogous for all calculators.

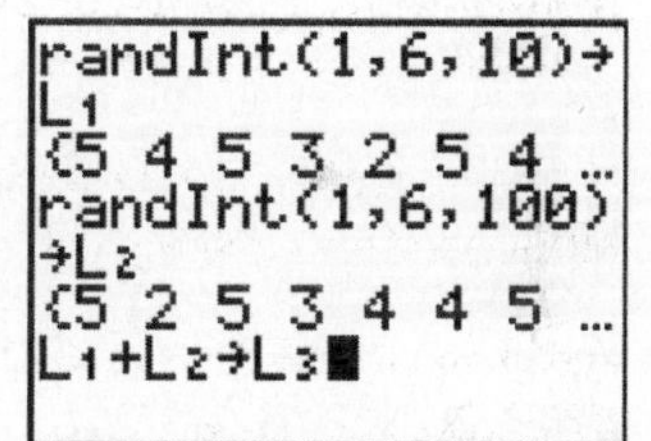

Generate 100 integers between 1 and 6 and store in `L1` (`list1`).
Generate 100 integers between 1 and 6 and store in `L2` (`list2`).
Add `L1` and `L2` and store the results in `L3`. On TI-83/84 calculators, this can either be done on the Home screen as at right, or by highlighting the new list name in the statistics editor and typing [2nd][1][+][2nd][2][ENTER] to execute the command `L1+L2`. On a TI-89, since the generation must be done within the statistics editor, it is easiest to perform the addition in the editor.

Press [STAT] [1] to see the results in your lists.

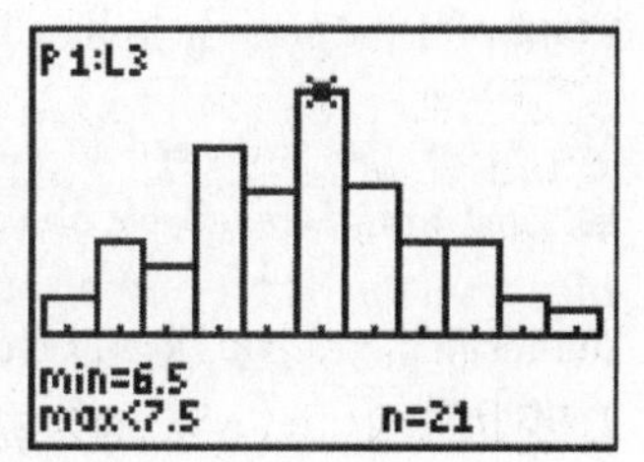

The screen at right shows a histogram of my results from this simulation. This histogram is for the data set consisting of the sums of the two rolls. This set was stored in `L3`. The Window settings were `Xmin` = 1.5, `Xmax` = 12.5, `Xscl` = 1, `Ymin` = -8, `Ymax` = 25. This simulation could be used to approximate the probability of rolling a certain sum. For example, we can see that our 100 rolls yielded 21 sums of 7. Thus we could estimate the probability of rolling a 7 to be 21%. (In actuality the value is 1/6 or 16.7%. More simulations would no doubt lead us closer to the truth.) Your results will vary.

COUNTING

In many probability problems, the big obstacle is trying to determine the number of possible (or favorable) outcomes. The procedures in this section can help.

Factorials

Notation: The factorial symbol is !

EXAMPLE Routes to Rides: How many different ride orders are possible if you want to visit Space Mountain, Tower of Terror, Rock 'n Roller Coaster, Mission Space and Dinosaur at Disney World on your first day?

By applying the factorial rule, we know that the 5 rides can be arranged in a total of 5! ways. We must calculate 5!. Do so on a TI-83 or -84, first type 5. Then press [MATH], arrow to PRB, and select option 4:!. Press [ENTER] to see the results of the calculation.

On a TI-89, the procedure is analogous, but keys are not the same, due to menu differences. On the home screen, type the 5 as above. Now, press [2nd][5] (Math), [7] (Probability), then [ENTER] to select option 1:! From the screen shown, press [ENTER] to transfer the factorial symbol to the input area, then [ENTER] again to complete the calculation.

On a TI-Nspire, locate the factorial (and permutations and combinations) by pressing [catalog] and locating them in the Probability section in the second menu section. Enter these on the Calculations screen similarly to the 89.

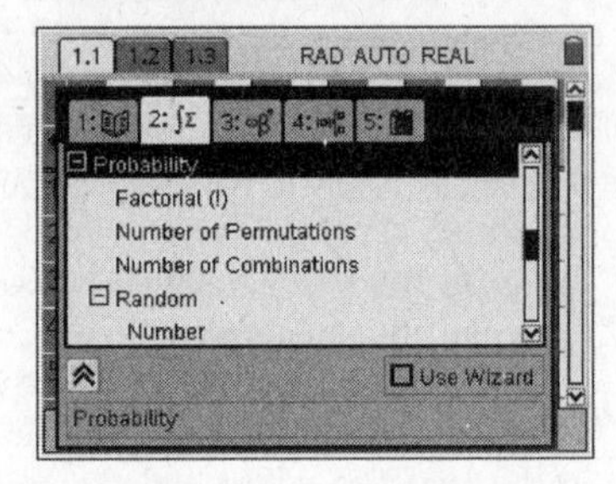

Permutations - All Items Distinct

EXAMPLE State Capitals: You want to conduct surveys in state capitals, but have only time enough to visit four of them. How many possible routes are there?

We know that we need to calculate the number of permutations of 4 objects selected from 50 available objects. Begin by typing 50. Press [MATH] [◄] [2] to select nPr from the Math, Probability submenu. Then type 4. Press [ENTER] to see the result 5,527,200. I think you need to add a few other conditions in planning your trip!

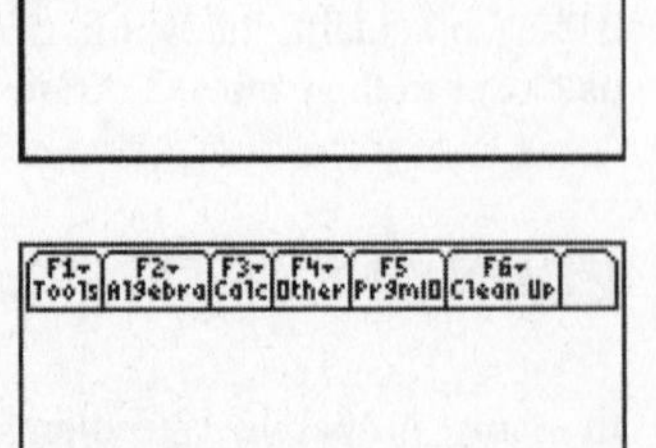

Permutations are also the second option on the Probability submenu on the TI-89 and Nspire, but one enters the command first, then the parameters n (the total to choose from) and k (the number desired) separated by a comma. Be sure to close the parentheses.

Permutations - All Items not Distinct

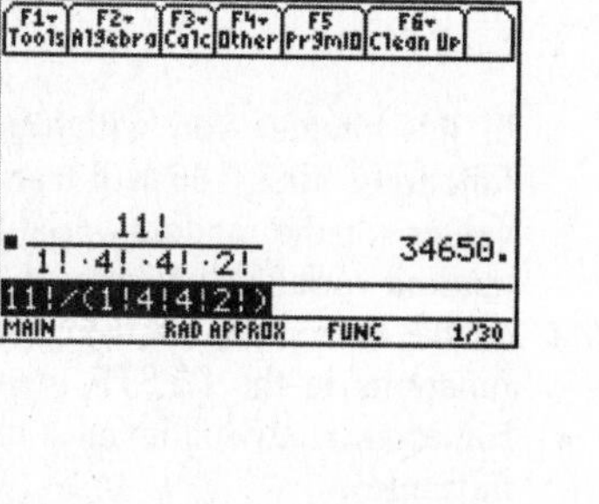

EXAMPLE Mississippi: How many distinct orderings are there of the letters in the word Mississippi? Questions like this are classics in the world of Probability. First, we note that there are 11 total letters in the word: 1-M, 4-i's, 4-s's, and 2-p's. Since one cannot (normally) distinguish one letter i from another, the total number of *distinct* permutations is $\frac{11!}{1!4!4!2!}$. My TI-89 calculation (TI-83 and -84 are analogous) shows there are 34,650 distinct combinations! That's a lot!

Combinations

EXAMPLE Phase I of a Clinical Trial: When testing a new drug on humans, a clinical test is normally done in three phases. Phase I is conducted with a relatively small number of healthy volunteers. Let's assume we want to treat 8 healthy humans with a new drug, and we have 10 suitable volunteers. If we want to treat all 8 patients at once, how many different treatment groups are possible?

10 nCr 8
45

We wish to select 8 from the available 10. But we are not concerned with the order of selection, merely the final result (the treatment group). On the home screen, type 10, then press [MATH], arrow to **PRB** and press [3] to select option `3:nCr`. Now type in 8 and press [ENTER]. We see there are 45 potentially distinct groups.

Maine Lottery: In the Maine lottery, a player wins or shares in the jackpot by selecting the correct 6-number combination when 6 different numbers from 1 through 42 are drawn. If a player selects one particular 6-number combination, find the probability of winning the jackpot. (Order is irrelevant).

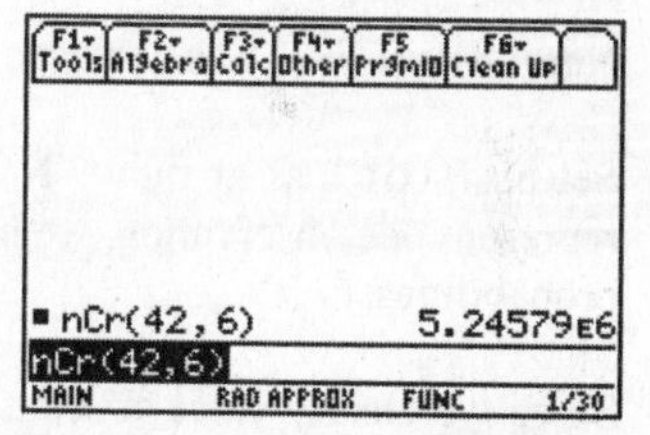

Because 6 different numbers are selected from 42 and order is irrelevant, we know we must calculate the number of combinations of 6 objects chosen from 42. This will tells us the total number of possible lottery outcomes. On a TI-83 or -84 we would do this as in the example above entering 42, then finding `nCr` and following that with a 6 before pressing [ENTER]. On a TI-89 home screen , we first locate the command, which is option 3 on the `Math, Probability` submenu. Press [2nd][5] (`Math`), [7] (`Probability`), [3] (`nCr`) to transfer the shell to the input area, then type 42,6) [ENTER] to complete the command. There are 5,245,790 possible winning combinations! No wonder lotteries are called a "tax on people who flunked statistics!"

WHAT CAN GO WRONG?

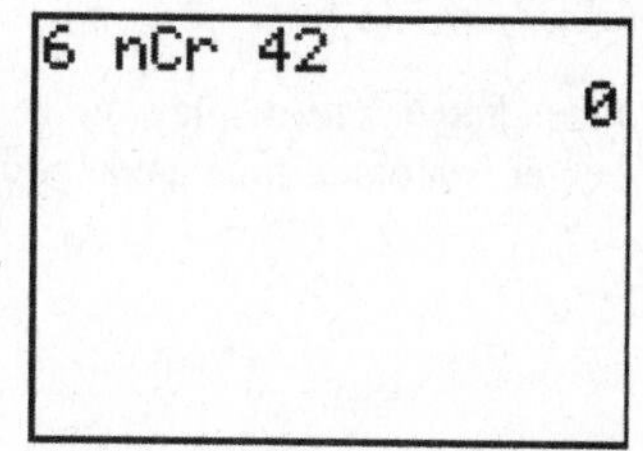

How can there be 0 combinations?
There can, but it's usually because you entered the parameters for the command backwards. One might be thinking of (in this example) choosing 6 from 42, but that's not how it works. The binomial coefficient (number of combinations) is read "n choose k." This helps keep straight that the total to choose from comes first.

5 Discrete Probability Distributions

In this chapter you will learn how your TI calculator can be used when working with probability distribution functions. First, you will learn how to use the calculator when given a probability function in the form of a table of values for the random variable with associated probabilities. You will also learn how to calculate probabilities of random variables which have the binomial or Poisson distribution. In both cases, you will be shown how to calculate on the Home screen using the formulas for these distributions and alternatively how to use built-in distribution functions in the DISTR menu (the Distributions section of the Probability menu on the TI-Nspire). Either alternative alleviates the need to use probability tables and, in fact, yields answers with more accuracy than the tables.

PROBABILITY DISTRIBUTIONS BY TABLES

EXAMPLE Mendel's Peas: Table 5-1, reproduced below, describes the probability distribution for the number of peas with green pods among 5 offspring peas.

X	0	1	2	3	4	5
P(x)	0.001	0.015	0.088	0.264	0.396	0.237

Probability Histogram

To plot the probability histogram from the table follow these steps:

Put the **X** values in L1 and the P(x) values in L2.

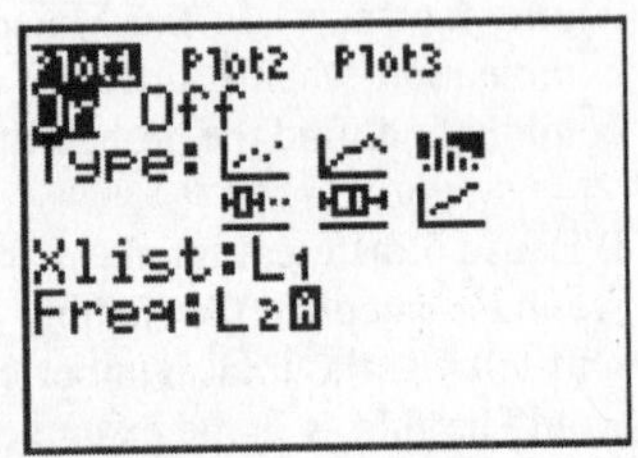

Note: Sum(L2) = 1 as expected.

Set up Plot1 as at right. Notice that since each value of the variable does not represent one observation, we have changed Freq: from 1 to L2 which contains the probabilities.

Since we clearly want to see a bar for each potential value of the variable X (the number of green pods), set the WINDOW values for this histogram as at right. Each bar has width 1, Ymax is set to .5 (slightly larger than the largest probability) and Ymin is set to -.1 (so TRACE information won't obscure the plot).

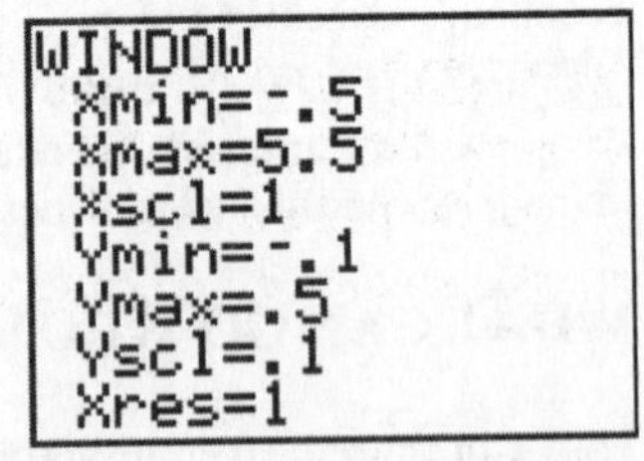

Note: These same WINDOW settings will work for a TI-89. Be sure to set "Use Freq and Categories" to Yes and specify the correct list of the frequencies.

Press GRAPH to display the plot. The distribution is left-skewed with a visual center is around three green pods. (This is the same graph as shown in Figure 5-3.)

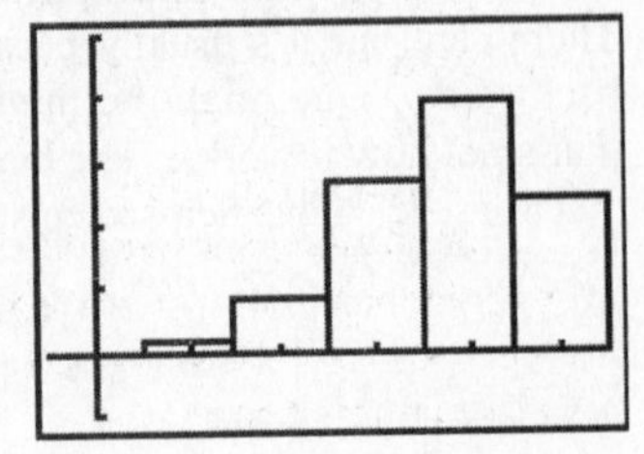

Nspire:

Enter the frequencies as integers (multiply by 1000), then use `freqTable▶List` from the Catalog to create a single list with the required number of repetitions of each value of the number of offspring peas. Insert a new Statistics graph page, and add the result variable (called peas here) as the X variable by pressing (menu), (2) for `Plot Properties`, then (enter) since option `4:Add X Variable` will be selected. Move the cursor to add the variable (that we called peas) as the X variable.

As usual, this plot will default to a dot plot. Change this to a histogram by pressing (menu), (1) for `Plot Type`, then (3) for `Histogram`. Finally, we want the plot in relative frequency. Press (menu), (2) for `Plot Properties`, and (enter) for option `2:Histogram Properties` that will be selected by default. Press (enter) to select option 1:Histogram Scale, then (3) for Density.

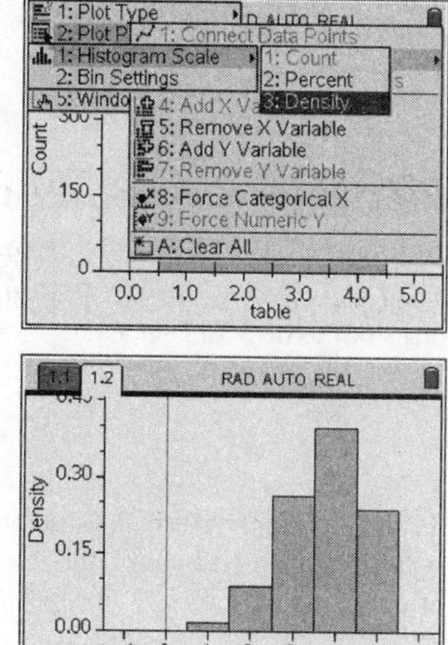

The finished plot is at right. Note that the first bar (with a height of 0.001) is not really visible; we do see a vertical line for the *Y* axis of the graph.

Mean, Variance and Standard Deviation

EXAMPLE Green Peas: Use the probability distribution to find the mean number green pea pods among five offspring peas, the variance, and the standard deviation.

As before, I have the **X** values in `L1` and the P(x) values in `L2`.

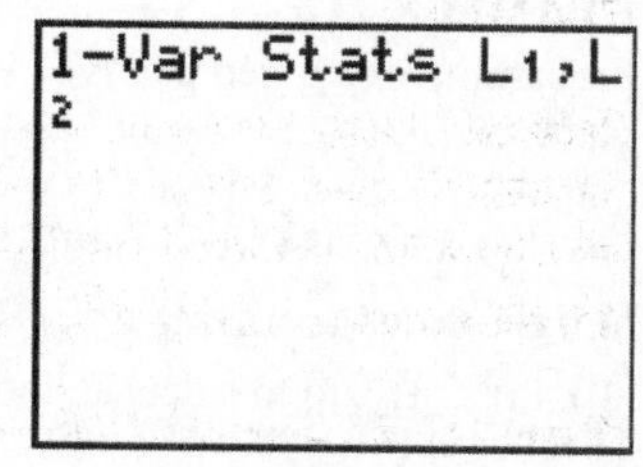

Press [STAT], arrow to `CALC` and press [ENTER] to select `1:1-Var Stats`. We must tell the calculator first which list has the variable (x) values, and then which list has the probabilities. Press [2nd] [1] [,] [2nd] [2]. Your command should look like the one at right. Press [ENTER] to see the results.

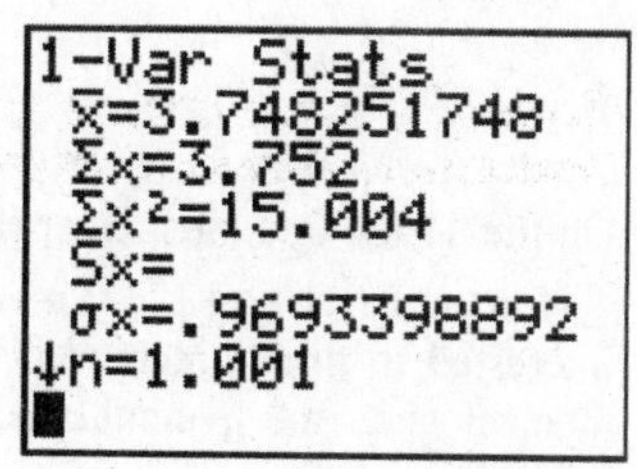

The value of the mean μ is given on this screen as $\bar{x} = 3.748$. The standard deviation σ is given by σx = 0.969, so the variance is the square of this, or $\sigma^2 = (0.969)^2 = 0.939$. The calculator is smart enough to recognize that you are dealing with a probability distribution, so only gives σx and not a sample standard deviation; however, it has only one symbol for the mean, so you must know this is a μ (population) mean. Note that we also see `n=1`. Use this as a double-check that frequencies have been properly input. If n is *not* 1, there is either some rounding error, or an input error.

TI-89: The procedure is analogous, but set `Use Freq and Categories` to `Yes`, and name the list of probabilities as the `Freq` list.

TI-Nspire: Use `One Variable Statistics` from the `Statistics, Stat Calculations` options in the spreadsheet application. Be sure to have named and specify the column of probabilities.

Expected Value/Mean by the Formula $\sum[x * P(x)]$

EXAMPLE Green Peas: Use the probability distribution to find the expected number or the mean number of green pea pods in five offspring.

Again put the **X** values in L1 and the P(x) values in L2. This is shown at right.

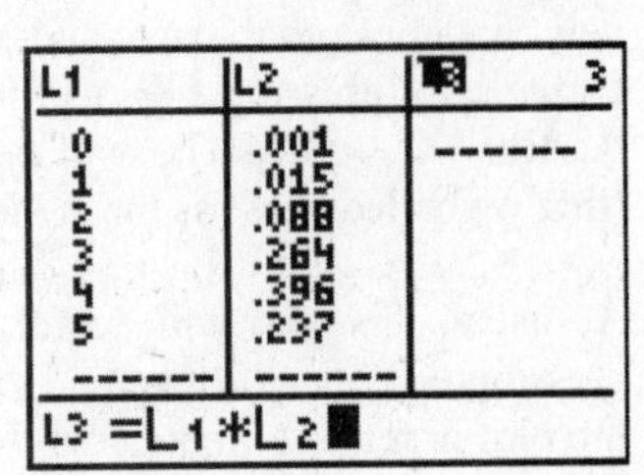

With list name L3 highlighted, multiply L1 by L2 as shown in the bottom line. Press [ENTER] to perform the calculation.

Press [2nd] [MODE] to Quit and return to the home screen. Then press [2nd] [STAT] [▶] [▶] [5] to paste the sum function from the LIST, MATH submenu on the home screen. Then press [2nd] [3] for L3. Press [ENTER] to find the sum of the entries in L3 which turns out to be 3.752 ≈ 3.8.

```
sum(L3)
                3.752
```

TI-89: Multiply the lists as above. To perform the sum operation, you can either place the cursor in a blank list within the statistics editor and use [F3] to access the List menu, arrow to Math, then select menu option 5:sum(or on the home screen, press [2nd][5] (Math), select option 3:List, then select option 6:sum(.

BINOMIAL DISTRIBUTION

Binomial Probability Formula
$P(x) = nCx*p^x(1 - p)^{n-x}$

EXAMPLE Green Peas: Find the probability of getting exactly 3 green pea pods in five offspring if the probability of a green pod is ¾ = 0.75. That is, find P(3) given that $n = 5$, $x = 7$, $p = 0.75$, and $q = 0.25$.

On a TI-83 or -84, type on the home screen to emulate what you see in the screen at right. You can find the nCr function in the MATH, PRB submenu (at [MATH] [◀] [3]). Press [ENTER] to calculate the probability as 0.264.

```
5 nCr 3*.75^3*.2
52
           .263671875
```

On the TI-89, one must enter the nCr function first. Press [2nd][5] for Math, then [7] for the Probability menu, then [3] for nCr. Out of five offspring, we are interested in three green pods, so complete the combination by entering 5,3 and then closing the parentheses. Complete the calculation by multiplying by *.75³*.25².

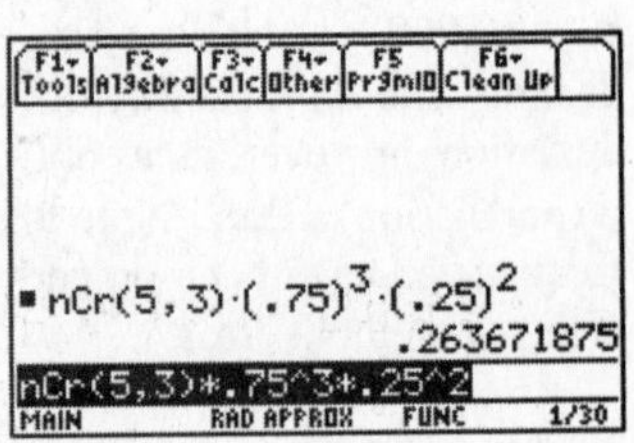

The procedure is analogous on the Nspire. Locate the Number of Combinations in the ⊞, then complete the rest of the calculation. Use the right arrow to move out of the exponents for more multiplication.

Built-in Binomial pdf and cdf Calculators

TI-83/84/Nspire Procedure

EXAMPLE Green Peas: We will extend the above example and show how the TI calculators' built-in functions for the binomial distribution can be used for a variety of problem types.

(a) Find the complete probability distribution.
To get the complete table of values and their probabilities, we first recognize that the possible number of green pea pods in five offspring is somewhere between 0 and 5. Fill list L1 with the values 0 to 5. With the list name highlighted, press [2nd][STAT] (LIST), arrow to OPS, then select option 5:seq(. Complete the command by typing X,X,0,5). Press [ENTER] to populate the list. To find the associated probabilities, move the cursor to highlight the name of list L2. Press [2nd] [VARS] (DISTR) and arrow to find binompdf. The actual location of this command will vary. It is menu option Ø on TI-83 calculators. If you have a TI-84 with operating system 2.30, it is menu option A. Then type 5 [,] 0.75 to specify that $n = 5$ and $p = 0.75$.

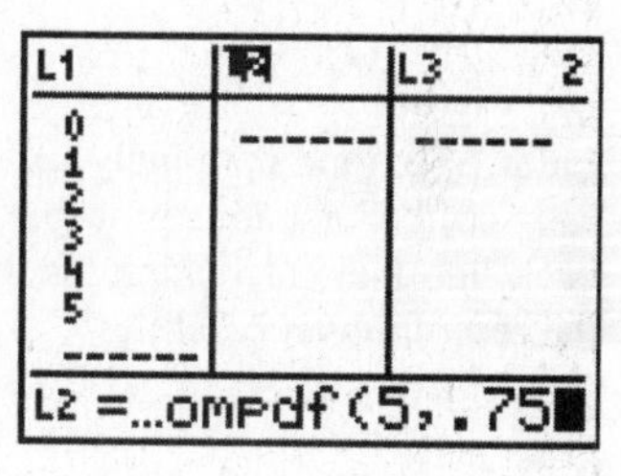

If you are using an Nspire, this is basically the same. Enter the command in the function area of a spreadsheet column, preceded by the = sign.

Press [ENTER] to calculate the probabilities. We can easily see the probabilities associated with different numbers of green pea pods in five offspring.

TI-89 Procedure:
In the Statistics Editor, press [F5] (Distr). Select option B:Binomial Pdf. In the dialog box, enter 5 for n, and 0.75 for p. Leave the box labeled x blank.

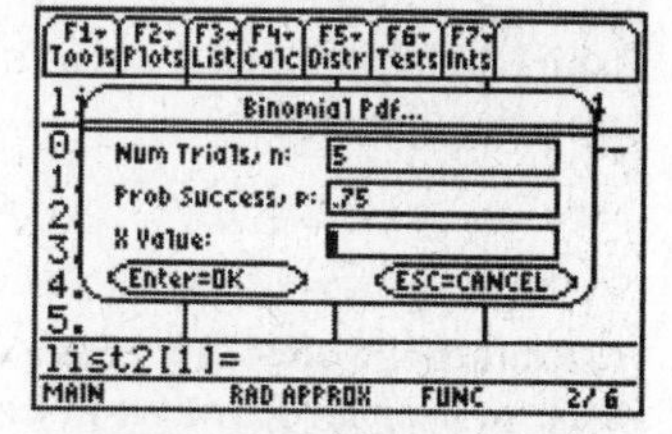

Pressing [ENTER] will display this results box.

Pressing ENTER again will display a new list labeled `Pdf` with the complete probability distribution. Please bear in mind that the first entry in the list is P(0), that is the probability of no green pea pods. As you scroll down the list of probabilities, subtract one from the subscript in the list for the appropriate value of **X**.

Nspire Procedure:
While in a Calculator screen, open the second (Catalog) section and locate and expand the `Probability, Distributions` segment. Locate the Binomial Pdf function toward the bottom of this block. Press (enter) to transfer the shell.

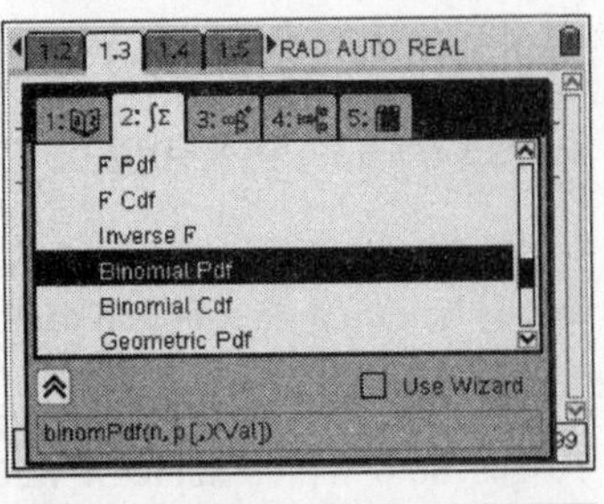

Enter `n`, `p`, (and optionally `x`) and press (enter) to perform the calculation. Here we can see the first four probabilities. Pressing the up arrow to move the cursor back into that calculation block, and then the right arrow will allow you to see the rest of the probabilities.

(b) Find the probability of getting *exactly* 3 green pods.
To find the probability at a single point, we use the `Binompdf` function. Press 2nd VARS, arrow to `Binompdf`, press ENTER to select it, then type 12 , .8 , 7 and then ENTER for the answer .0532 as seen in this screen. Of course, the answer agrees with what was seen on the previous page when we used the formula explicitly.
Note: pdf stands for probability density function

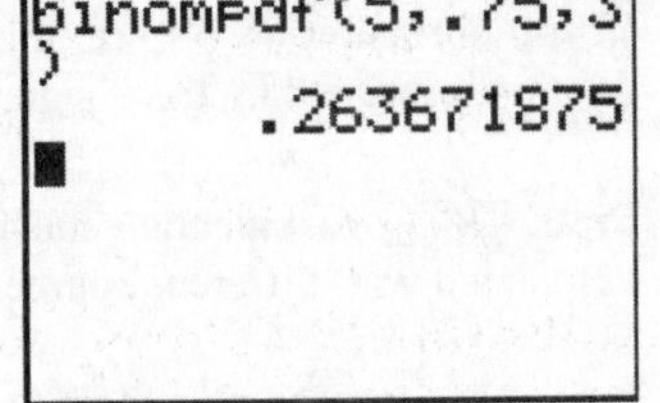

TI-89/Nspire: Use `Binomial Pdf` as described above, except now specify the `x` value of 3. Pressing ENTER in the dialog box displays the results at right.

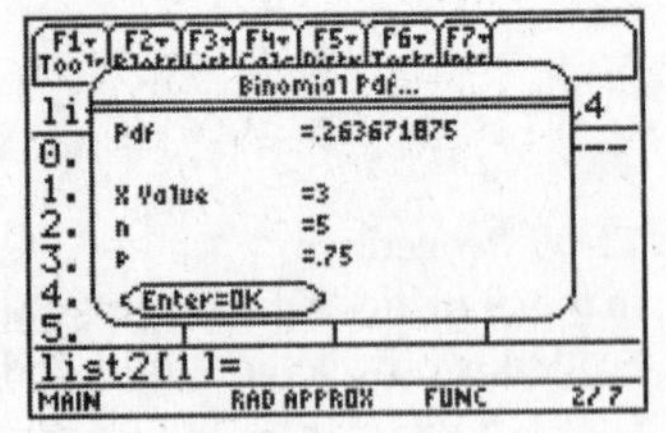

Nspire: This command can be used to fill a single cell in the spreadsheet, or in the calculator application.

(c) Find the probability of *at most* three green pods. (TI-83/84)
At most means "less than or equal to", so the probability of at most 3 is in fact P(0)+P(1)+P(2)+P(3). This can be found by using the built-in `binomcdf` (`Binomial Cdf` on an 89) function. This is right below `Binompdf` on the `DISTR` menu (option A, B, or C depending on calculator model. Hint: locate it by pressing ▲ to go to the bottom of the menu and then up from there.) With the command pasted on the home screen (or in the 89 dialog box), type 12,0.8,7 and press ENTER. The answer is 0.367.

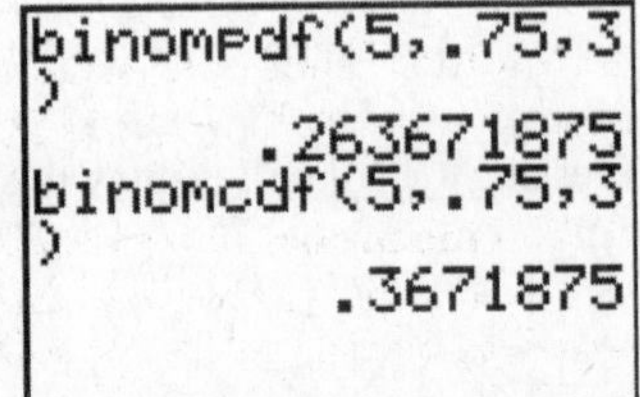

(d) Find the probability of *at least* 3 green pods. (TI-83/84)
At least means "greater than or equal to". We want P(3)+P(4)+P(5). Since the `Binomcdf` function finds $P(X \leq k)$, we will find this as 1 - [P(0)+P(1)+P(2)].

```
        .263671875
binomcdf(5,.75,3
)
         .3671875
1-binomcdf(5,.75
,2)
        .896484375
```

(e) *At most* and *at least* (TI-89 and Nspire)
The `Binomial Cdf` function on these calculators works differently. In both cases, one specifies the `Lower Value` (the lowest x value to be included) and the `Upper Value` (the largest x value to be included). The calculation set up in this screen will find $P(\mathbf{X} \leq 3)$. To find $P(\mathbf{X} \geq 3)$, one specifies a lower value of 3 and an upper value of 5.

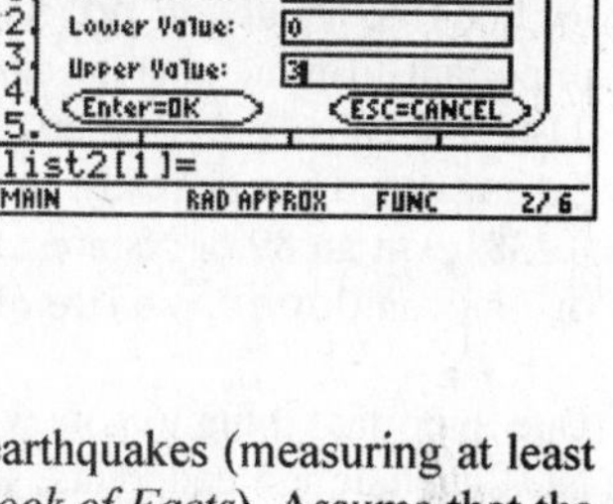

POISSON DISTRIBUTIONS

EXAMPLE Earthquakes: For a recent period of 100 years, there were 93 major earthquakes (measuring at least 6.0 on the Richter scale) in the world (based on data from the *World Almanac and Book of Facts*). Assume that the Poisson distribution is a suitable model. If so, we expect 93/100 earthquakes per year (on average).

Poisson Probability Formula $P(x) = \mu^x e^{-\mu}/x!$

(a) Find the probability there are no major earthquakes in a year.
Recalling that in this problem $\mu = 0.93$, and noting that for this question $x = 0$, we evaluate the above formula on the home screen as seen at right. **Note:** to get e^(, one presses [2nd][LN] on an 83 or 84. On an 89, e^(is [♦][X]. Also remember that on an 83 or 84, factorials are menu choice 4 from [MATH], PRB; on an 89 they are choice 1 from the `Math, Probability` menu.

```
.93^0*e^(-.93)/0
!
      .3945537104
```

Built-in Poisson pdf and cdf Calculators

(a) Find the probability of exactly two quakes in a year.
We can use the built-in function `poissonpdf(` to find the probability of a Poisson variable being equal to some given value x. This function is located in the `DISTR` menu toward the bottom (actual location depends on calculator model). Paste on the function and fill in the rest. Press [ENTER] to see the same result as before. Note that the `poissonpdf(` function requires two inputs: μ and x. In the screen below, we see this probability is 0.171, or 17.1%.

```
poissonpdf(.93,2
)
      .1706247521
poissoncdf(.93,2
)
      .9321134131
```

(b) Find the probability of *at most* two major quakes.
The built in function `poissoncdf(` can be used to find cumulative probabilities of a Poisson variable up to and including a given value x. This function is located immediately below poissonpdf on the `DIST` menu. Paste it on the main screen and fill in the rest. We see the cumulative probability is .9323. Note the inputs for the `poissoncdf` function were the same as for the pdf function, μ and x. This probability is 0.932, or 93.2%.

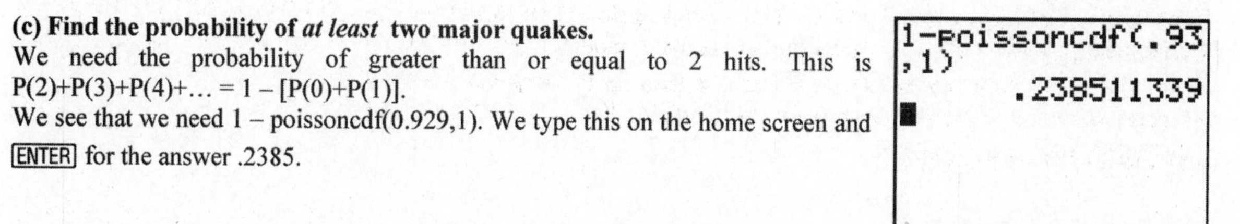

(c) Find the probability of *at least* two major quakes.
We need the probability of greater than or equal to 2 hits. This is P(2)+P(3)+P(4)+… = 1 – [P(0)+P(1)].
We see that we need 1 – poissoncdf(0.929,1). We type this on the home screen and [ENTER] for the answer .2385.

```
1-poissoncdf(.93
,1)
       .238511339
```

To find this probability on a TI-89 (or Nspire), remember that on these models we specify the low end of interest and the high end of interest (strictly, infinity or ∞, but practically a large number will do). In this dialog box I have specified the `Upper Value` (high end of interest) as 1000, since practically speaking, there is no probability above here. After pressing [ENTER] we get the same answer.

(d) Find the probability of between 2 and 6 major quakes, inclusive.

We see that we need P(2)+P(3)+P(4)+P(5)+P(6). We note that this will require us to find the value of `poissoncdf(0.929,6) – poissoncdf(0.929,1)`. It is important to remember to close the parentheses after `poissoncdf(0.929,6` on TI-83/84 calculators, otherwise it will not know that you are done with that `poissoncdf` and ready to move to another one. We find this probability is 0.2381. On an 89 or Nspire, simply use poissoncdf and specify a `Lower Value` of 2 and an `Upper Value` of 6.

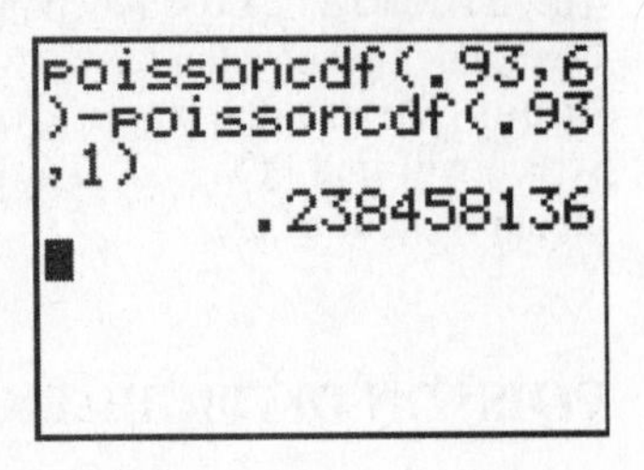

One important thing to note when dealing with discrete distributions that a question reading "between a and b" does NOT include the endpoints, since mathematically this would be P(a < **X** < b). This is different from specifically including the endpoints, as above.

WHAT CAN GO WRONG?

Err: Domain?

This error is normally caused in these types of problems by specifying a probability as a number greater than 1 (in percent possibly instead of a decimal) or a value for *n* or *x* which is not an integer. Reenter the command giving *p* in decimal form. Pressing [ESC] will return you to the input screen to correct the error. This will also occur in older TI-83 calculators if *n* is too large in a binomial calculation; if that is the case, you need to use the normal approximation.

How can the probability be more than 1?

It can't. If a probability looks more than 1 on the first glance, check the right hand side. This value is 9.7×10^{-18} or seventeen zeros followed by the leading 9.

6 Normal Probability Distributions

In this chapter you will use your TI calculator to aid you in several types of problems dealing with the normal Probability Distribution. First, you will learn to use the built-in function `normalcdf` to obtain probabilities associated with normal random variables. You will also learn to use the built-in function `invNorm` to obtain percentiles for a given normal distribution (that is, when given a probability value you will find the associated value of the normal variable). Both of these functions are on the `DISTR` menu at [2nd] [VARS] ([F5] `Distr` on an 89, or in the `Probability Distributions` section of the Catalog on an Nspire). Using them will alleviate the need to use the tables in the text. We will do some calculations that illustrate the Central Limit Theorem and how to approximate a binomial distribution with a normal distribution. You will learn how to obtain a *normal quantile plot.* This plot, described in the text, is used to determine if a given set of data might come from a population with a normal distribution.

FINDING PROBABILITIES FOR A NORMAL RANDOM VARIABLE

We will use the built-in function `normalcdf(`. This function is in the `DISTR` menu as option 2. ([2nd] [VARS] [2] on TI-83 or 84 calculators). It requires 4 inputs. They are (a,b,μ,σ). Here the "a" and "b" denote two values between which you want the probability. As you might guess, μ and σ represent the mean and standard deviation of the normal variable. (If we do not put in values for μ and σ, the calculator assumes the distribution is *standard* normal and thus $\mu = 0$ and $\sigma = 1$). We will use the function on several examples to follow.

Standard Normal Probabilities

EXAMPLE Scientific Thermometers: The Precision Scientific Instrument company manufactures thermometers that are supposed to give readings of 0° C at the freezing point of water. Tests on a large sample of these instruments reveal that at the freezing point of water, some thermometers give readings above 0° C (denoted by positive numbers) and some give readings below 0° C (denoted by negative numbers. If we assume that the mean reading is 0° C, and the standard deviation of the readings is 1° C, and that the readings are normally distributed (have a bell-shaped distribution), then these thermometers have a *standard normal distribution,* because the mean is 0 and the standard deviation is 1.

(a) If a thermometer is randomly selected, find the probability that its reading at freezing will be *less than* 1.27°.

TI-83/84 Procedure:

Press [2nd] [VARS] [2] to paste `normalcdf(` onto the home screen. Then type –E99,1.27 and [ENTER]. To get the "E" press [2nd] [,].We find the answer is 89.8%; there is a 89.8% probability a random thermometer will read less than 1.27°.

Note: Normal Distributions are (theoretically) defined on the entire real line $(-\infty, \infty)$. The value –E99 $(-1\text{x}10^{99})$ is a very large negative number (that the TI-83/84 calculators understand as $-\infty$. We used it in the "a" position because we had no specific value at which to start our probability area; we wished to get all the area <u>below</u> 1.27. In practice, however, a very "large negative" number (say –999999 could be used for $-\infty$. Note also as shown in the screen that we did not have to specify the mean and standard deviation for a *standard* normal probability problem.

TI-89 Procedure:

In the Statistics/List Editor application, press [F5] (`Dist`) and select menu option `4:Normal Cdf`. The dialog box asks for the `Lower Value` of interest (−∞ since we want all possible values less than 1.27), the `Upper Value` of interest (1.27), the mean (0), and the standard deviation (1) for the distribution of interest. To enter −∞, press [♦][X].

Press [ENTER] for the results. We see there is a 89.8% chance a randomly selected thermometer of this type will read less than 1.27°C at the freezing point of water.

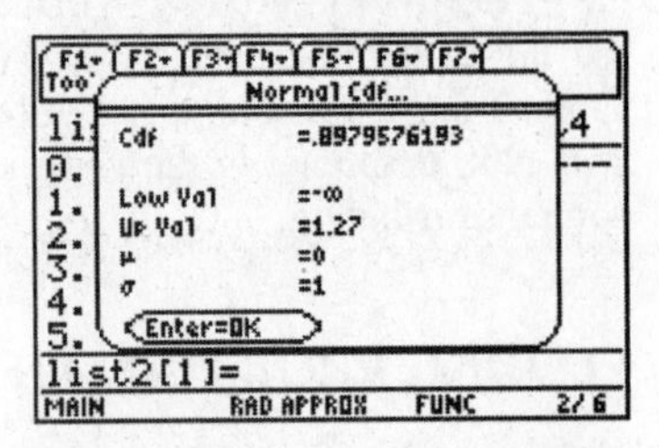

Nspire Procedure:

After locating the command in the `Probability Distributions` segment of the Catalog, enter the bounds. Note that with this calculator, we must specify the lower end as −1E99.

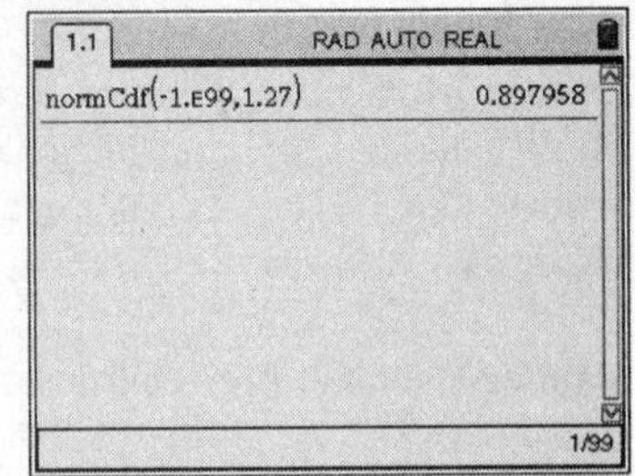

(b) Find the probability of a reading *above* –1.23°.

We proceed similarly to problem (a) but with different values for "a" and "b." We get 0.8907.

Note: This time we used E99 in the "b" position because it was a "greater than" problem and we had no specific value for the upper end of our probability area.

(c) Find the probability of a reading *between* –2° and 1.5°.

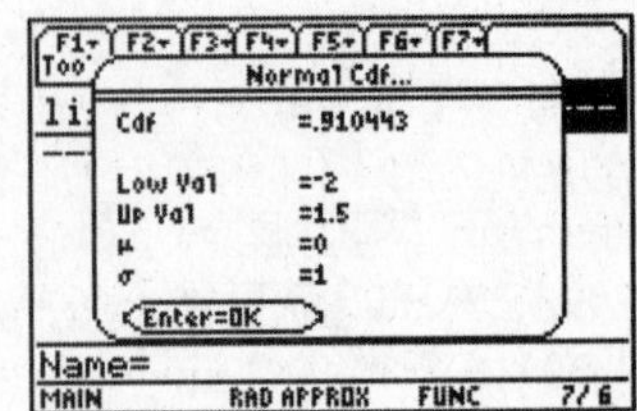

This is very straight-forward because in *between* problems, we are given values for inputs "a" and "b." We see the probability of a reading between -2° and 1.5° is about 91%.

Non-Standard Normal Probabilities

There is really no difference in the procedure for standard and non-standard normal probabilities using TI calculators except for the fact that in the non-standard normal problems the mean and standard deviation are required inputs for the `normalcdf` function.

EXAMPLE Doorway Heights: The typical home doorway has a height of 6 ft 8 in, or 80 inches. Because men tend to be taller than women, we will consider only men as we investigate the limitations of that standard doorway height. Given that heights of men are normally distributed with mean 69.0 in. and standard deviation 2.8 in., find the percentage of men who can fit through the standard doorway without bending or bumping their heads.

We use normalcdf to answer the question, specifying a low end (mathematically) of −E99, a high end of 80 (inches), mean 69, and standard deviation
2.8. We see that over 99.99% of men should be able to fit through these doors.

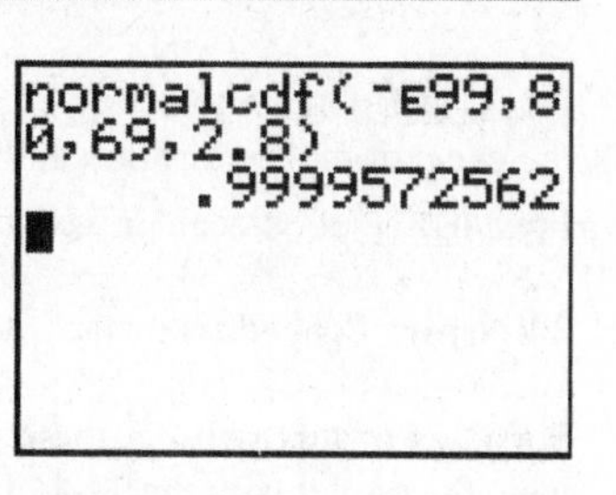

EXAMPLE Birth Weights: Birth weights in the United States are normally distributed with a mean of 3420 g and standard deviation 495 g. The Newport General Hospital requires special treatment for babies that are less than 2450 g (unusually light) or more than 4390 g (unusually heavy). What percent of babies do not require special treatment? These babies weigh between 2450 and 4390 g. We see in the screen at right that about 95% do not require special treatment.

SKETCHING NORMAL AREAS

Your text suggests sketching the curve and shading the area corresponding to what is desired. This is very good advice, as it can help one decide if the values obtained from a calculation seem reasonable (Were all the parameters of the calculation input correctly?) TI calculators can also sketch normal curves and shade the desired area.

Caution: Before using this procedure, first turn off all StatPlots.

EXAMPLE: IQ Scores: A psychologist is designing an experiment to test the effectiveness of a new training program for airport security screeners. She wants to begin with a homogeneous group of subjects having IQ scores between 85 and 125. Given that IQ scores are normally distributed with a mean of 100 and a standard deviation of 15, what percentage of people have IQ scores between 85 and 125?

TI-83/84 Procedure:

Setting the WINDOW.
We have used the following criteria to set our window values:
Xmin = $\mu - 3\sigma$= 100 – 3(15) = 55
Xmax = $\mu + 3\sigma$ = 100 + 3(15) = 145
Xscl and Yscl = 0
Ymin = -.1/ σ = -.1/15 = -0.01
Ymax = 0.4/ σ = 0.4/15 = 0.03

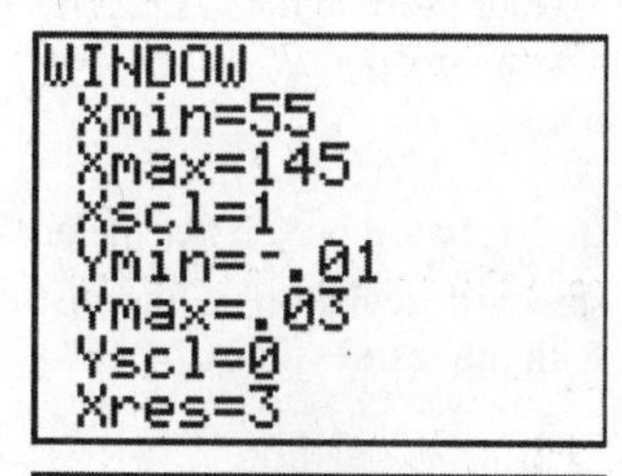

Getting the Graph: Press [2nd] [VARS] [▸] [1] to get ShadeNorm(pasted on your home screen. Fill in the values of a, b, μ and σ as usual. Press [ENTER] for the shaded. We can see the Area =.7936. This area is equivalent to the probability we were seeking.
Caution: If repeatedly using this procedure, the normal curve may become totally shaded! Between uses, press [2nd][PRGM] (Draw) and press [ENTER] to select 1:ClrDraw. About 79.4% of all people should have IQ scores between 85 and 125.

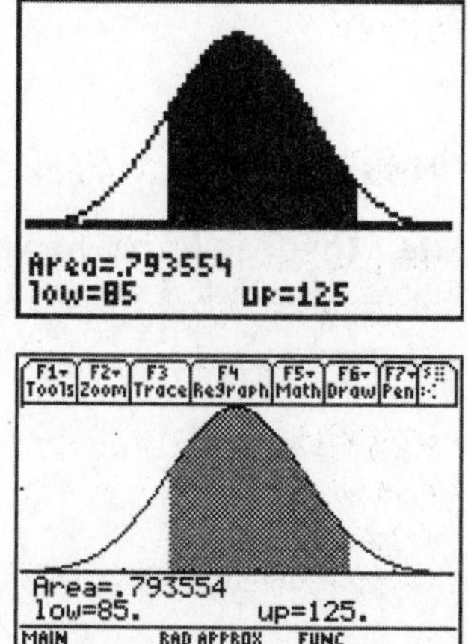

TI-89 Procedure:

There is an option in the TI-89 dialog box which makes setting the window automatic. Press [F5] (Distr) and press ⊙ to expand the 1:Shade submenu.

Press [ENTER] again to select `1:Shade Normal`. The input dialog box looks just like that for `Normal Cdf`, with the exception of one option at the bottom: `Auto-Scale`. Press ▶ to expand the choices, and set this option to Yes. Press [ENTER] to execute the command.

TI-Nspire Procedure:

Sorry, as of this writing, these calculators will not create these graphs. If you really want to produce one and you have the model with the TI-84 faceplate, change that and use TI-84 mode.

FINDING VALUES OF A NORMAL R. V. (INVERSE NORMAL PROBLEMS)

The InvNorm Function

EXAMPLE Scientific Thermometers: Using the same thermometers as earlier, find the temperature corresponding to P_{95}, the 95th percentile. That is, find the temperature separating the bottom 95% from the top 5%.

TI-83/84 Procedure:

Press [2nd] [VARS] [3] to paste `invnorm(` onto your home screen. This function needs three inputs: p, μ, and σ. Here p represents the percentile we desire to know (the area *to the left* of the desired point on the distribution). So for this problem we input 0.95. (We need not input μ, and σ since the thermometers are standard normal) Press [ENTER]. We see that 95% of these thermometers will read 1.64° or less at the freezing point of water.

TI-89 Procedure:

The TI-89 has the capability of computing inverses for several distributions. In the Statistics List Editor, press [F5] (`Distr`) then arrow to `2:Inverse` and press ▶ to expand the submenu. Press [ENTER] to select `1:Inverse Normal`.

In the dialog box, enter the area to the left of the point desired, then the mean and standard deviation of the distribution. Pressing [ENTER] displays the output box, with the result value of 1.64°.

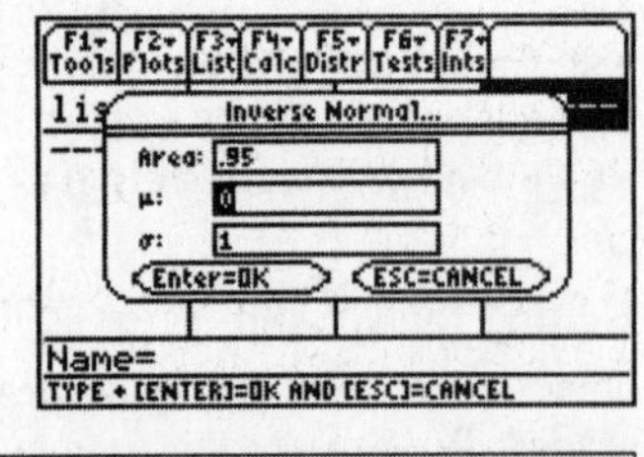

Nspire Procedure:

Select `Inverse Normal` from the Catalog's Probability Distributions section. Enter the parameters similarly to those for the 83/84 series.

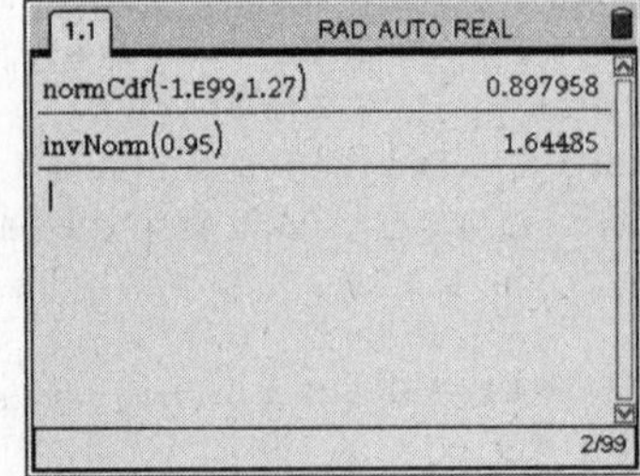

EXAMPLE Designing Door Heights: When designing a door, how high should it be if 95% of men will fit through without bending or bumping their head? Heights of men are normally distributed with a mean of 69.0 in, and a standard deviation of 2.8 in. We see in the screen at right that the 95th percentile of men's heights is 73.6 in (6 ft. 1.6 in.).

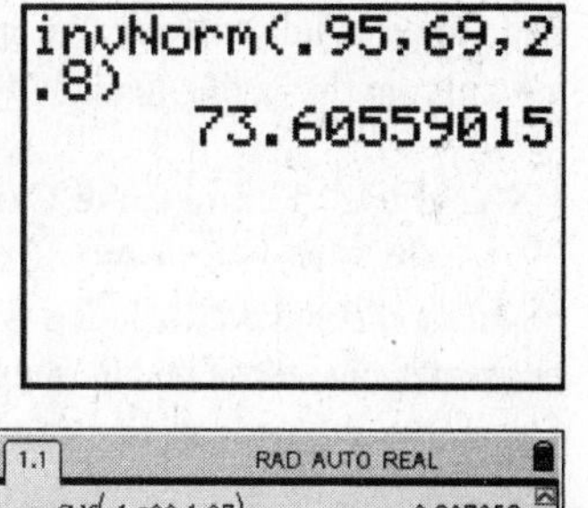

EXAMPLE: Birth Weights: The Newport General Hospital wants to redefine the minimum and maximum birth weights that require special treatment. After considering several factors, a committee recommends special treatment for birth weights in the lowest 3% and highest 1%. Find the birth weights that separate the lowest 3% and highest 1% if birth weights are normally distributed with a mean of 3420 g and a standard deviation of 495 g.

```
1.1                       RAD AUTO REAL
normCdf(-1.E99,1.27)           0.897958
invNorm(0.95)                   1.64485
invNorm(0.03,3420,495)          2489.01
invNorm(0.99,3420,495)          4571.54
|
                                   4/99
```

We see at right that the lowest 3% of birth weights are those less than 2489 g. The highest 1% (there are 99% less than this weight) are about 4572 g and higher. Note that this answer differs slightly due to rounding errors from that in the text.

CENTRAL LIMIT THEOREM

As the sample size increases the sampling distribution of the sample means approaches a normal distribution.

EXAMPLE Water Taxi Safety: Assume (based on the National Health and Nutrition Examination) that the population of men have weights that are normally distributed with a mean of 172 lb and a standard deviation of 29 lb.

(a) Find the probability that if an individual man is randomly selected, his weight will be greater than 174 pounds. The probability is 0.4725.

(b) Find the probability that 20 randomly selected men will have a mean weight that is greater than 174 pounds (so the total weight exceeds the safe limit of 3500 pounds for the water taxi.) The probability is 0.3789.

```
normalcdf(174,E9
9,172,29)
        .472508463
normalcdf(174,E9
9,172,29/√(20))
        .3788802417
```

Note that the only thing that has changed is the standard error which is now $29/\sqrt{20}$. You can enter it explicitly as I have in this example, or perform the calculation first ($29/\sqrt{20} = 6.484597...$) but the temptation to round too much may be tempting! In any case, we can see that the chance of overloading the water taxi (having total weight more than 3500 lb) with 20 randomly selected men is too high. There should be a lower person limit for safety's sake.

Note: In the above, our answers differ slightly from those in the text because we did not use the tables. The calculator solutions are actually more precise because they do not suffer from round-off error.

EXAMPLE Body Temperatures: If we assume that humans have a mean body temperature of 98.6ºF with standard deviation 0.62 ºF, what is the chance of getting a mean reading of 98.2 ºF or lower from a random sample of 106 people?

We don't know the shape of the original distribution, but with the large sample size, the sample mean should have a normal distribution due to the Central Limit Theorem. Therefore, we can answer the question using `normalcdf` as at right. We need to be careful reading the answer. At first glance, the probability might look larger than 1 (1.552). *But, probabilities can't be larger than 1!* Look at the right-hand portion of the answer: `E-11`. This means the leading digit (1) occurs in the eleventh decimal place, so the answer is really 0.000000000016. A probability

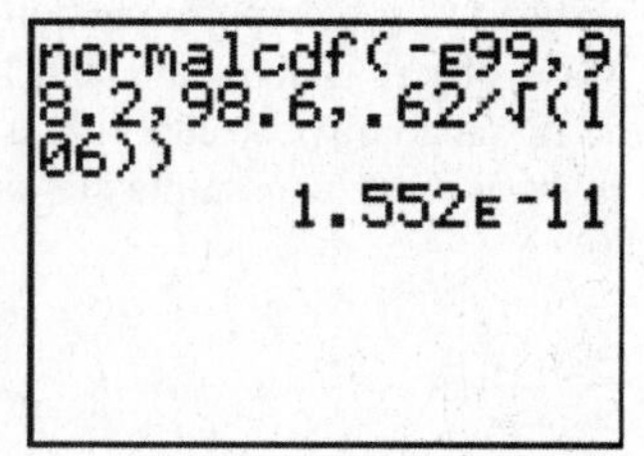

this small usually means that given an observation (the sample mean actually seen) something is wrong with an assumption (here, the mean body temperature is probably less than the 98.6 ºF usually assumed.)

EXAMPLE Filling Coke Cans: Cans of regular Coke are labeled to indicate that they contain 12 oz. A sample of n = 36 cans had mean $\bar{x}$ = 12.19 oz. If the Coke cans are filled so that $\mu = 12.00$ oz (as labeled) and the population standard deviation is $\sigma = 0.11$ oz, find the probability that a sample of 36 cans will have a mean of 12.19 or greater. Do these results suggest that the Coke cans are filled with an amount greater than 12 oz?

```
normalcdf(12.19,
E99,12,.11/√(36)
)
   1.85385254E-25
```

We answered the question using `normalcdf`. According to our calculation, the chance a sample of 36 cans would have a mean this high (or higher) is $1.9\text{x}10^{-25}$. This small probability indicates that cans are actually filled with a mean amount more than 12 oz.

NORMAL APPROXIMATION OF A BINOMIAL DISTRIBUTION

If np ≥ 5 and n(1-p) ≥ 5, then the binomial random variable is approximately normally distributed with the mean and standard deviation given by $\mu = np$ and $\sigma = \sqrt{np(1-p)} = \sqrt{npq}$. (Actually, the criterion varies from textbook to textbook, but this is what Triola uses.)

EXAMPLE Mail Survey: Dr. Triola was mailed a survey from Viking River Cruises, and the survey included a request for an e-mail address. Assume that the survey was sent to 40,000 people and that for such surveys, the percentage of responses with an e-mail address is 3%. If the true goal of the survey is to acquire a bank of at least 1150 e-mail addresses, find the probability of getting at least 1150 responses with e-mail addresses.

The mean number of responses with e-mail addresses is $\mu = 40000 * 0.03 = 1200$, and the standard deviation is $\sigma = \sqrt{40000 * .03 * .97} = 34.11744$. We use `normalcdf` to find the desired probability (of at least 1150 responses with e-mail addresses) using the continuity correction to consider that 1150 or more includes all the area to the right of 1149.5. We find the probability of at least 1150 responses with e-mail addresses is 93.06%.

EXAMPLE Internet Use: A recent survey showed that among 2822 randomly selected adults, 2060 (or 73%) stated that they are Internet users (data from Pew Research Center). If the proportion of all adults using the Internet is actually 0.75, find the probability that a random ample of 2822 adults will result in exactly 2060 users.

```
normalcdf(2059.5
,2060.5,2116.5,2
3.002717)
   8.496360827E-4
binompdf(2822,.0
3,2060)
                0
```

Again, we check to see if we can use the normal approximation. The mean is np = 2822*.75 = 2116.5 > 5 and we also have $n(1-p)$ = 2822*.25 = 705.5 > 5. We need to find the standard deviation σ = $\sqrt{2822(.75)(1-.75)} = 23.002717$. According to the continuity correction, we will consider "exactly 2060 Internet users" to be anywhere between 2059.5.5 and 2060.5 Internet users. Notice we get results using the approximation and the binomial (exact) model that are extremely close. There is about a 0.0008% chance to find exactly 2060 Internet users among 2822 randomly selected adults (actually, finding exactly *any* single number of Internet users in a sample this size would be small).

NORMAL QUANTILE PLOTS

Normal quantile plots are a good means of assessing whether (or not) a set of data comes from an (approximately) normal distribution. Unlike histograms or boxplots, they can be constructed for small or large sets of data. If the distribution is normal, the plot will be approximately a straight line.

EXAMPLE Movie Lengths: The first five movie lengths in Data Set 9 of Appendix B are 110, 96, 170, 125, and 119 minutes. A histogram or boxplot will not be useful in assessing normality for these data. Construct a normal quantile plot and determine whether they appear to come from a normally distributed population.

TI-83/84 Procedure:

With the data values entered in `L1`, we define the plot using the last plot type on the `STAT PLOTS` screen. Note that we can select to have the data axis be on either the X or Y axis; your text puts data on the X axis.

Press [ZOOM][9] to display the plot. These points (except for possibly the last) are roughly on a line. These data may be normally distributed.

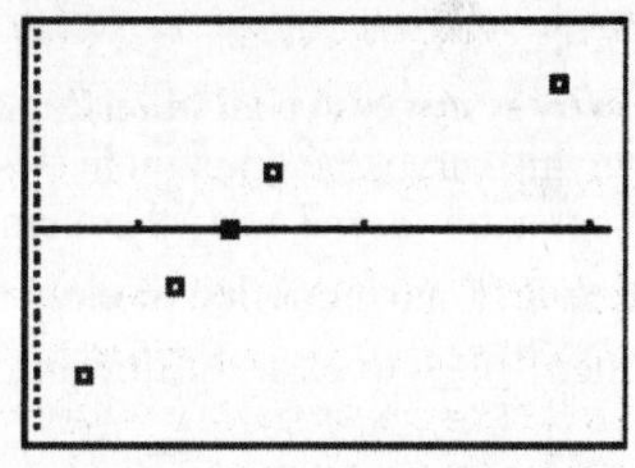

TI-89 Procedure:

The data have been entered in `list1`. Creating this plot on these calculators involves an additional step that computes the z-scores for each data value (as if the data were normally distributed).

Press [F2] for the `Plots` menu, then [2] for `Normal Prob Plot`.

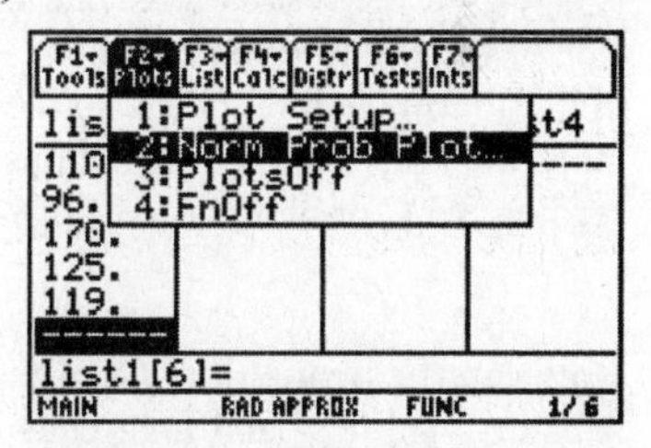

The plot number defaults to one more than what is currently defined. Select the appropriate list name, which axis has the data (your text uses the X axis), and a mark for each data point. Notice that *z*-scores will be stored for you in the `statvars` folder. Press [ENTER] to calculate the *z*-scores.

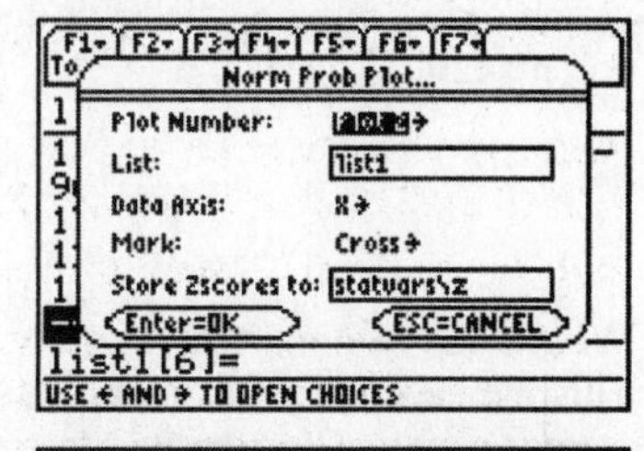

Return to the [F2] `Plots` menu, make certain that any other plots are unchecked (use [F4] to uncheck them) and press [F5] for `ZoomData`. These points (except for possibly the last) are roughly on a line. These data may be normally distributed.

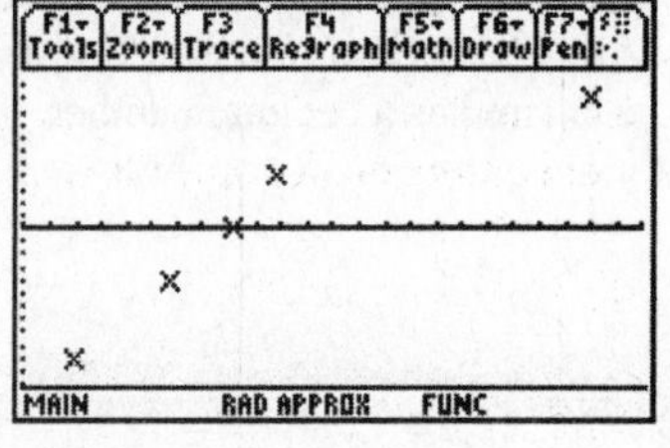

Nspire Procedure:

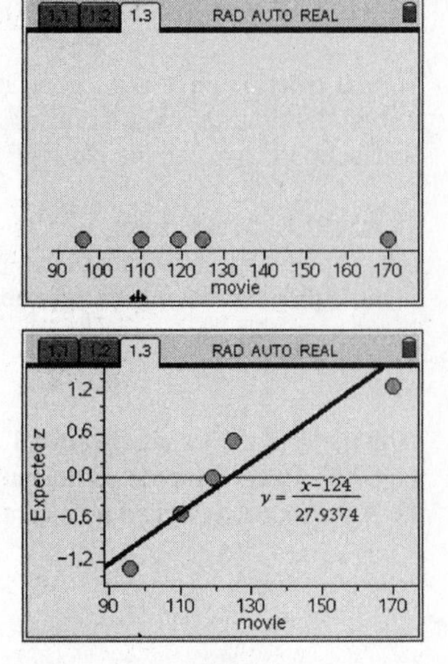

The data have been entered in a spreadsheet column we have named `movie`. Press (home)(5) to add a statistics plot page to the current document. Move the cursor to the bottom of the graph, and click to add `movie` to the graph. At this point, we have a dot plot of the data.

Press (menu), (1) for `Plot Type`, and (4) for `Normal Probability Plot`. Note that the Nspire adds the straight line for reference (making it easier to see that all these points are close to a straight line), and shows the calculation of the *z*-scores.

WHAT CAN GO WRONG?

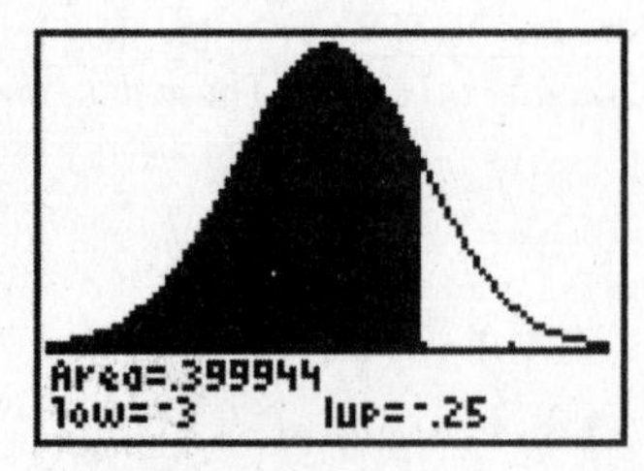

Why is my curve all black?
In this curve, the graph indicates more than half of the area is of interest between z-scores of –3 and -.25; the message at the bottom says the area is 40%. This is a result of having failed to clear the drawing between commands. Press [2nd][PRGM] then [ENTER] to clear the drawing, then reexecute the command.

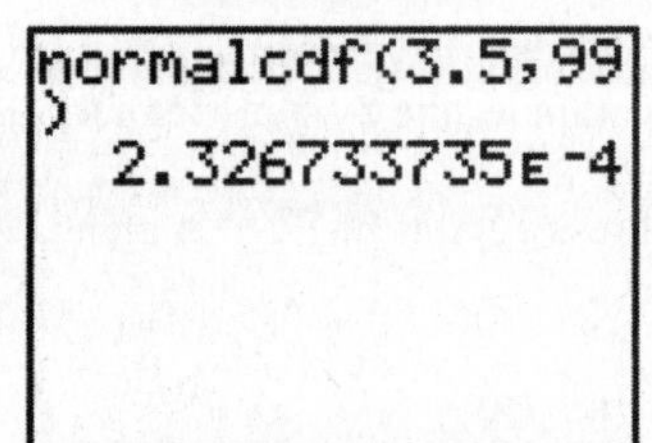

How can the probability be more than 1?
It can't. If the results look like the probability is more than one, check the right side of the result for an exponent. Here it is –4. That means the leading 2 is really in the fourth decimal place, so the probability is 0.0002. The chance a variable is more than 3.5 standard deviations above the mean (this would be a thermometer reading 3.5°C or more at freezing) is about 0.02%.

How can the probability be negative?
It can't. The low and high ends of the area of interest have been entered in the wrong order. As the calculator does a numerical integration to find the answer, it doesn't care. You should.

What's Err: Domain?
This message comes as a result of having entered the `invNorm` command with parameter 90. (You wanted to find the value that separates the top 10% of IQ scores, so 90% of the area is to the left of the desired value.) The percentage must be entered as a decimal number. Reenter the command with parameter .90.

7 Estimates and Sample Size

In this chapter you will learn how your TI calculator can aid you in estimating population parameters using the results of a single random sample. In this work, you will primarily use some of the options on the STAT TESTS menu (Ints on an 89). You will learn how to estimate population proportions, means and variances as well as how to estimate the sizes of the samples you will need in each setting.

As usual, the first time a significant difference between the TI-83/84, TI-89, or Nspire calculators is encountered, we present all procedures. After an initial run-through with a confidence interval, the differences are clearly apparent.

ESTIMATING A POPULATION PROPORTION

EXAMPLE A Global Warming Poll: In the Chapter Problem, a Pew Research poll of 1501 adults had 70% answer "yes" to the question "From what you've read and heard, is there solid evidence that the average temperature of earth has been getting warmer over the past few decades, or not?" We wish to estimate the true population proportion. Find the margin of error E that corresponds to a 95% confidence level for the proportion. Then find the 95% confidence interval

TI-83/84 Procedure:

Press STAT, arrow to TESTS, arrow down the menu and choose option A (actually, it is more efficient to use the *up* arrow, or press ALPHA MATH = A). You will see a screen like that at right. Note that you need 3 inputs: x, n and C-level. In this problem it is obvious that n = 1501 and C-level = .95. It is <u>important</u> to understand that the x which is needed is the <u>number</u> of responses with the characteristic of interest (<u>not</u> the percentage or proportion). Sometimes you are given x, but sometimes you are given the sample proportion $\hat{p}$. To find x when you have been given the sample proportion $\hat{p}$ (as in this case) you must use the formula $x = n\hat{p}$ and round to the nearest integer. Here, we have x = 0.70*1501 = 1050.7 (round this to 1051). You can actually do the multiplication on the input line for x – press ENTER to do it, then arrow back up and do any needed rounding.

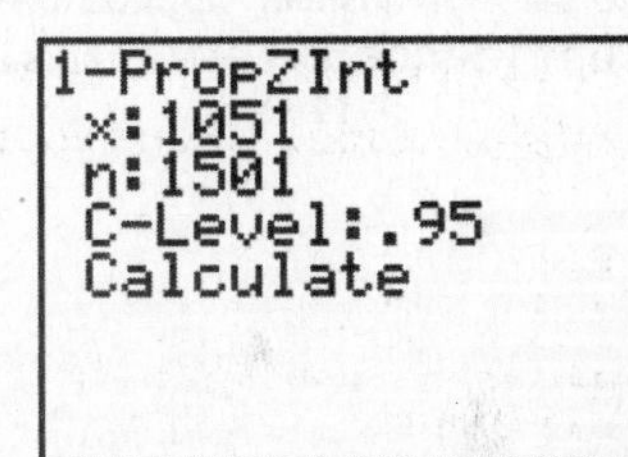

Note: The value of x <u>must</u> be an integer or else your TI calculator will have a domain error.

Highlight the word Calculate and press ENTER to see the results. The interval is given in parentheses. It is (0.67702, 0.72338). We normally report the interval only to the nearest 1/10th of a percent; here we would report our 95% confidence interval as 0.677 to 0.723, or 67.7% to 72.3%. Based on this poll, between 67.7% and 72.3% of all adults believe the earth has been getting warmer over the past few decades, with 95% confidence.

```
1-PropZInt
 (.67702,.72338)
 p̂=.7001998668
 n=1501
```

If you wish to find the margin of error E which was used to calculate this interval, you can find the difference of the two endpoints of the interval and divide by 2, or subtract $\hat{p}$ from the upper end. We find that the margin of error was $E = .023$.

```
(.723-.677)/2
                .023
.723-.700
                .023
```

TI-89 Procedure

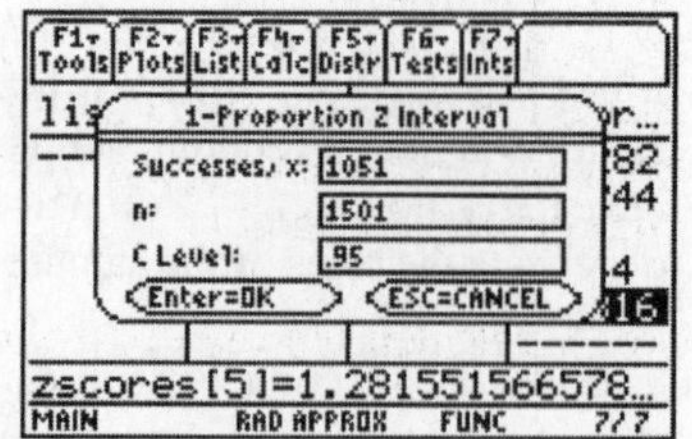

From the Statistics List Editor, press [2nd][F2] = [F7] (Ints). 1-PropZInt is option 5 on this menu. After selecting the menu option (either press [5] or scroll down to highlight the option and press [ENTER]) you will seen an input dialog box like the one at right.

Input the number of successes, x, (this must be an integer), the number of trials, n, and the desired Confidence Level. If you are given the sample results as a proportion (percentage) of successes, multiply by n and round to find x. You cannot do the multiplication .7*1501 on this screen; do it on the home screen and then enter the result here.

Press [ENTER] to display the results. Notice that the TI-89 gives the margin of error (ME) as well as the ends of the interval.

Nspire Procedure

In the spreadsheet application, press (menu), (4) for Statistics, (3) for Confidence Intervals, and (5) for 1-Prop Z Interval.

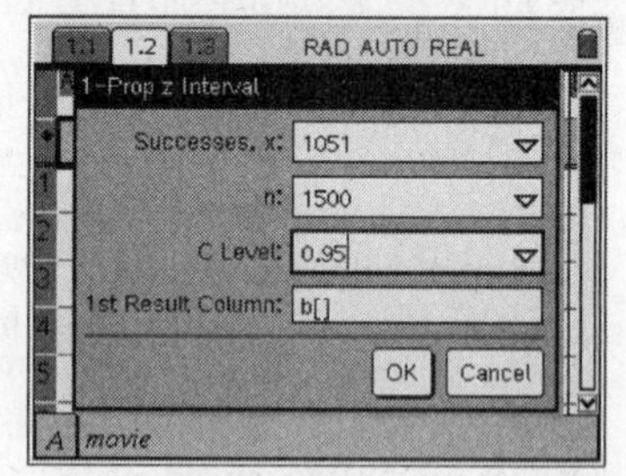

As with the TI-89, any multiplication must be done on a calculations screen (you can't do it here). Use the right arrow to progress through the boxes. Note that the default is to have results begin in the first empty column of the spreadsheet. You can change this, but it's usually unnecessary. Press (enter) to do the calculations when all the boxes have been filled in.

The results are shown at right. This calculator also gives the margin of error (ME) explicitly. If you want to see more digits, as always, with the cursor in the column, you can press (menu), (1) for Actions, (2) for Resize, and select either (2) Maximize Column Width or (1) Resize Column Width. If you select option 1, pressing the right arrow will make the column wider; pressing the left arrow will make the column narrower.

Home Screen Calculation for Proportion Estimates

Using the same example, you could calculate the margin of error and the confidence interval for the population proportion on your home screen.

You must find the critical value $Z_{\alpha/2}$ for a 95% confidence level. To find this value, calculate as follows: α = 1 - 0.95 = 0.05, α/2 = 0.05/2 = 0.025. This means we need the value which separates the top 2.5% from the bottom 97.5% of the standard normal distribution (or the bottom 2.5% from the top 97.5% because of symmetry). We use the `invNorm` function as described in Chapter 6. We see the value needed is 1.96.

```
invNorm(.025
         -1.959963986
```

Now to calculate the value of the margin of error, *E*, use the formula $E = z_{\alpha/2}\sqrt{\frac{\hat{p}\hat{q}}{n}}$. We know from the problem that $\hat{p} = 0.70$, $\hat{q} = 0.30$ and $n = 1501$. This is shown at the top of the screen at right. We have stored the margin of error as E, indicated with the → symbol.

Now we continue to calculate the lower and upper endpoints of the confidence interval by taking our sample proportion $\hat{p} = 0.70$ and subtracting and adding E. We obtain nearly the same two interval endpoint values found when we used `1-PropZInt` (the differences are due to rounding).

```
1.96√(.7*.3/1501
)→E
          .0231833063
.7-E
          .6768166937
.7+E
          .7231833063
```

Determining Sample Size Required to Estimate the Population Proportion

The formula for sample size is not built into TI calculators. We calculate it on the home screen. It is $n = \frac{[z_{\alpha/2}]^2 \hat{p}\hat{q}}{E^2}$, where $\hat{p} = 0.5$ if no other estimate is available.

EXAMPLE How Many Adults Use the Internet?: Suppose a manager for E-Bay wants to determine the current percentage of U.S. adults who use the Internet. How many adults must be surveyed in order to be 95% confident that the sample percentage is in error by no more than three percentage points? a) Use the result from an earlier Pew study which said 73% of U.S. adults used the Internet in 2006. b) Assume that we have no prior information suggesting a possible value of $\hat{p}$.

First find the critical z-value for 95% confidence. As explained above, we use `InvNorm` with $\alpha/2 = 0.025$. We find the value is 1.96. Ignore the negative – the normal distribution is symmetric, and *z* will be squared.

We show the calculations using both values for $\hat{p}$. We find that the sample size needs to be 842 if we use the prior estimate for $\hat{p}$ and 1068 if we do not have a prior estimate and must use $\hat{p}$ = 0.5. Notice that having a "guessed" starting value can save lots of work!

```
1.96²*.73*.27/.0
3²
             841.3104
1.96²*.5*.5/.03²
          1067.111111
```

ESTIMATING A POPULATION MEAN: σ KNOWN

The formula for the confidence interval used to estimate the mean of a population when the standard deviation is known is built into the calculator. It is option 7 on the `STAT TESTS` menu on the TI-83/84 (option 1 on the `Ints` menu on a TI-89 or Nspire) and is called `ZInterval`. This built-in function can be used with either raw data that has been entered in a list or summary statistics. The interval could also be calculated on the home screen in a similar manner as was shown for the confidence interval for a population proportion. The procedure for all calculators is similar.

EXAMPLE Weights of Men: People have died in boat and aircraft accidents because an obsolete estimate of the mean weight of men was used. In recent decades the mean weight of men has increased considerably, so we need to update our estimate of that mean so that boats, aircraft, elevators, and other such devices do not become dangerously overloaded. Using the weights of men from Data Set 1 in Appendix B, we obtain these sample statistics for the simple random sample: $n = 40$ and $\bar{x} = 172.55$ lb. Research from several other sources suggests that the population of weights of man has a standard deviation given by $\sigma = 26$ lb. Construct a 95% confidence interval estimate of the mean weight of all men. What do the results say about the mean weight of 166.3 lb that was used to determine the safe passenger capacity of water vessels in 1960?

ZInterval with Summary Statistics

```
ZInterval
 Inpt:Data Stats
 σ:26
 x̄:172.55
 n:40
 C-Level:.95
 Calculate
```

Press STAT, arrow to TESTS and select option 7:ZInterval (option 1 on the TI-89 or Nspire Ints menu) to get a screen similar to the one at right. Your first choice is the type of Input you will use. For this problem you have summary statistics (not raw data entered in a list), so you should move the cursor to highlight Stats and press ENTER. This choice is presented on the TI-89 and Nspire before the input dialog box is shown. Now fill in all the other information which was given in your problem.

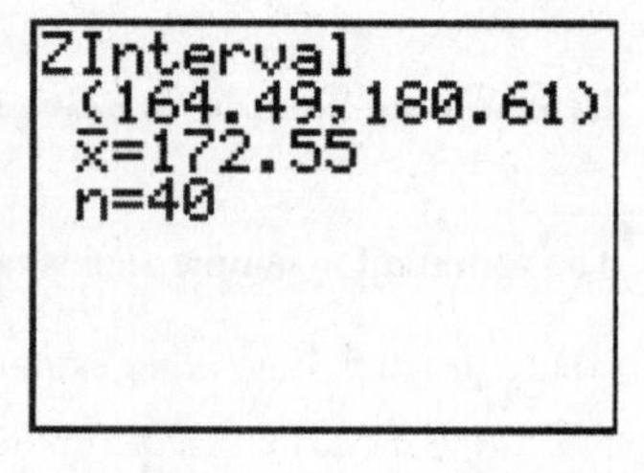

Highlight Calculate and press ENTER to display the results. Based on this sample, we are 95% confident the average weight of a man is between 164.5and 180.6 lb. Note that the 1960 value of 166.3 lb is included in the interval; it is *possible* the mean weight of a man is still 166.3 lb; however, considering that this is at the lower end of the interval (and the dangerous consequences of overloading), we'd prefer to either collect more data (or be safe and increase the "average weight" standard).

```
n=40

(180.61-164.49)/
2
                8.06
```

To retrieve the margin of error that was used to build the interval, we again must find the difference between the upper and lower endpoints of the interval and divide this difference by 2. The margin of error was 8.06 lb. (The TI-89 and Nspire again give the margin of error (ME) explicitly in their results).

ZInterval with Raw Data

Exercise 27 gives the blood pressure levels of 14 different second-year medical students at Bellevue Hospital. If we assume that the population standard deviation of blood pressure measurements is 10 mmHg, what is a 95% confidence interval for the mean blood pressure reading of second-year medical students at this hospital? For convenience, we repeat the measurements:

138	130	135	140	120	125	120	130	130	144	143	140	130	150

Enter the data into list L1. (Do this by typing it in or transferring from another source.)

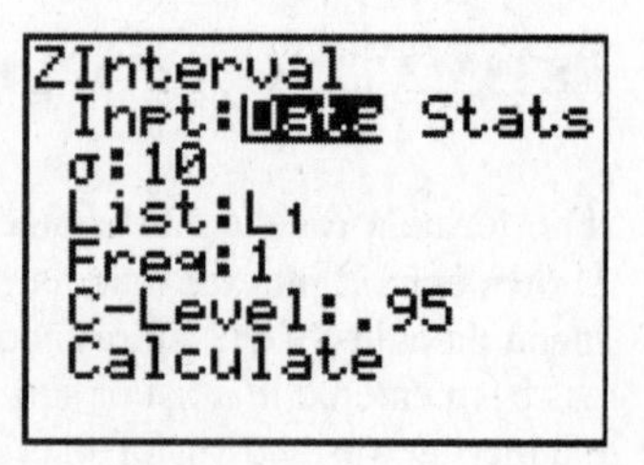

Press STAT, arrow to TESTS and select ZInterval. Your first choice is the type of Input you will use. For this problem you have raw data, so you should move the cursor to highlight Data and press ENTER. Now fill in all the other information. Note you must specify the standard deviation (because it is supposed to be known) and also the list where you have stored your data.

Highlight the word `Calculate` and press ENTER. Based on this sample information, we estimate the mean blood pressure reading for second-year medical students at this hospital (we don't have data representative of *all* second-year medical students) is between 128.7 and 139.2 mmHg.

Home Screen Calculations for Mean Estimates

To calculate the margin of error and confidence interval for the men's weight example on the home screen, we would first need to know the value of $Z_{\alpha/2}$ for a 95% confidence level. We eralier found this in to be 1.96 using `invNorm`. Here, we used the formula $\bar{x} \pm E$ where $E = z_{\alpha/2} \frac{\sigma}{\sqrt{n}}$. We first calculated E, stored it as E and then used it to find the endpoints of the confidence interval for the mean weight. Again, the interval is 164.5 to 180.6 lb.

Determining Sample Size Required to Estimate μ

TI calculators do not have a built-in function to calculate sample size. The formula can be calculated on the home screen. The procedure is similar to the previous example of a sample size calculation for a proportion problem, but the formula is different: $n = \left[\frac{z_{\alpha/2}\sigma}{E}\right]^2$.

EXAMPLE: IQ Scores of Statistics Students: Assume that we want to estimate the mean IQ score for the population of statistics students. How many statistics students must be randomly selected for IQ tests if we want 95% confidence that the sample mean is within 3 IQ points of the population mean (assuming σ = 15)?

We have found before that for 95% confidence, $Z_{\alpha/2} = 1.96$. We also have σ = 15, and $E = 2$. Putting these together with the formula (be careful with parentheses!) tells us we must sample at least 97 statistics students to get the desired margin of error (remember, always "round" up!)

```
(1.96*15/3)²
                96.04
```

ESTIMATING A POPULATION MEAN: σ NOT KNOWN

The formula for the confidence interval used to estimate the mean of a population when the standard deviation σ is not known is also built into the calculators and is called `TInterval` (Option 8 on the STAT, TESTS menu of TI-83/84 calculators, option 2 on the `Ints` menu on the TI-89 or Nspire.) This built-in function can be used with either raw data or summary statistics just like the `ZInterval` already discussed. The interval could also be calculated on the home screen in a similar manner as was shown for the other confidence intervals in this section.

We first illustrate finding the critical value $t_{\alpha/2}$. The `InvT (Inverse t)` function is built into the TI-84, -89, and Nspire; if you are using a TI-83, you will need to use tables to find this (but with the built-in functions, this is usually not necessary).

EXAMPLE: Finding a Critical t Value: A sample of size $n = 7$ is a simple random sample selected from a normally distributed population. Find the critical value $t_{\alpha/2}$ corresponding to a 95% confidence interval.

Just as with finding critical value from a standard normal distribution, t distributions are symmetric around 0, so we can find the value with 2.5% to the left of $t_{\alpha/2}$ (and ignore the negative sign) or find the value with 97.5% to the left. Use the `InvT` function on a TI-84 (`Inverse t` on an -89) from the `DISTR` menu. To find this on the Nspire, use `Statistics, Distributions` either in a spreadsheet or calculations window. The parameters are the area to the left of the desired point and the degrees of freedom (one less than the sample size with a single sample). Note that this critical value is larger than the corresponding z value. With small samples, there is more uncertainty that needs to be reflected in our answers.

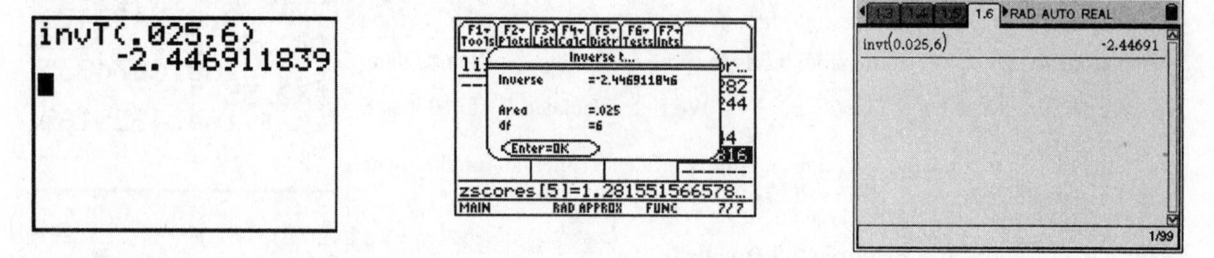

EXAMPLE Garlic for Reducing Cholesterol: A common claim is that garlic lowers cholesterol levels. In a test of the effectiveness of garlic, 49 subjects were treated with doses of raw garlic, and their cholesterol levels were measured before and after the treatment. The changes in their levels of LDL cholesterol (in mg/dl) have a mean of 0.4 and a standard deviation of 21.0. Use the sample statistics to construct a 95% confidence interval for μ, the mean net change in LDL cholesterol after the garlic treatment. What does the confidence interval suggest about the effectiveness of garlic in reducing LDL cholesterol?

TInterval with Summary Statistics

Press STAT ▶ ▶ 8 to get the input screen for a `TInterval` (this is option 2 on the `Ints` menu for the TI-89 and Nspire). Your first choice (just as with the `ZInterval`) is the type of Input you will use. For this problem you have summary statistics (not raw data), so you should highlight on `Stats` and press ENTER. Now fill in all the other information which was given in the problem statement as you see it on the screen.

```
TInterval
 Inpt:Data Stats
 x̄:.4
 Sx:21
 n:49
 C-Level:.95
 Calculate
```

Highlight the word `Calculate` and press ENTER for the results. We are 95% confident the mean change in LDL cholesterol level with raw garlic is between -5.6 and 6.4 mg/dl. Since this interval contains 0 (no net change) it is possible that raw garlic has no effect on LDL cholesterol.

```
TInterval
 (-5.632,6.4319)
 x̄=.4
 Sx=21
 n=49
```

To retrieve the margin of error that was used to build the interval, we again must find the difference between the upper and lower endpoints and divide this difference by 2. The margin of error was 6.0.

We see below that the procedure on the Nspire is similar; this can be done in either a spreadsheet or calculations window. Note the margin of error is given as ME.

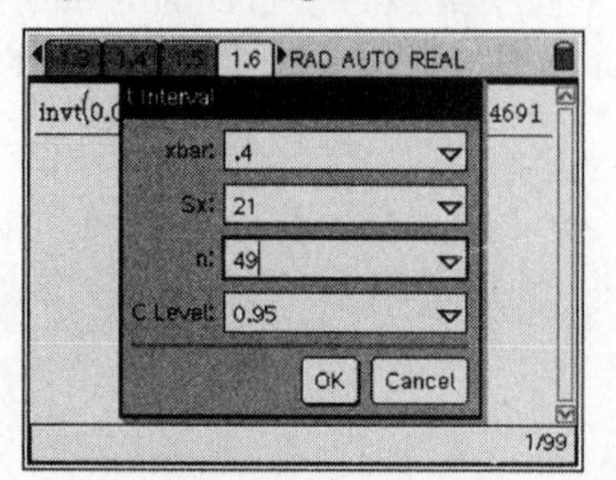


```
"CLower"      -5.6319
"CUpper"       6.4319
"x̄"            0.4
"ME"           6.0319
"df"           48.
"sx := sn-1x"  21.
"n"            49.
```

TInterval with Raw Data

EXAMPLE Constructing a Confidence Interval: The data below are the ages of applicants who were unsuccessful in winning promotion (Barry and Boland, *American Statistician,* Vol. 58, No. 2)

34	37	37	41	42	43	44
45	45	45	46	48	49	53
53	54	54	55	56	57	60

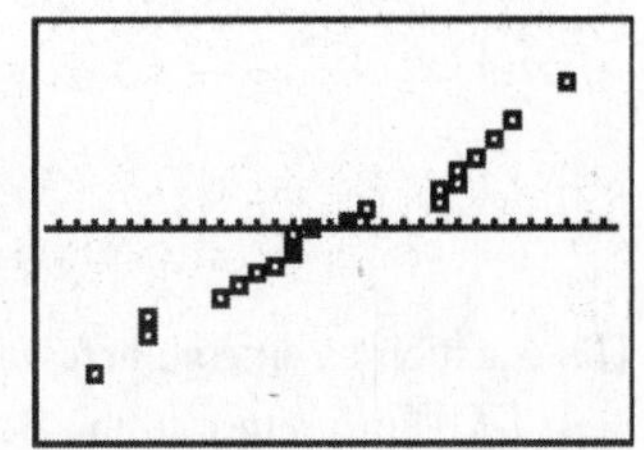

Put the data into list L1. (Do this by typing it in or transferring from another source.) Because this is a small sample ($n < 30$), we should check to make sure it is approximately normal (no outliers or strong skewness). Looking at the data values, there do not appear to be any unreasonably large or small values. A normal plot of the data also shows the data are approximately normally distributed.

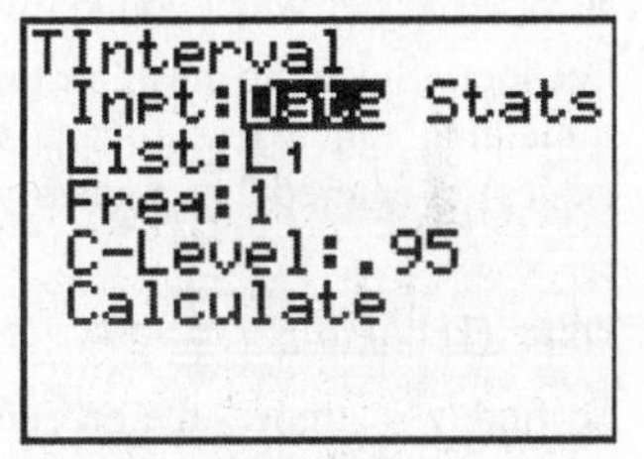

Press STAT, arrow to TESTS and select option 8 (Ints and option 2 on an 89) to get a screen for the TInterval. Your first choice is the type of Input you will use. For this problem you have raw data, so you should move the blinking cursor to highlight Data and press ENTER to move the highlight. Now fill in all the other information. This time you are not asked to specify σ because it is supposed to be unknown when using the TInterval (the calculator will find $\bar{x}$ and s for you). You must, of course, specify the list where you have stored the data.

```
TInterval
 (44.224,50.824)
 x̄=47.52380952
 Sx=7.249958949
 n=21
```

Highlight the word Calculate and press ENTER. Based on this sample, the average age of those who would be denied promotion in the scenario described is between 44.2 and 50.8 years old, with 95% confidence.

ESTIMATING A POPULATION VARIANCE

Methods for computing confidence intervals for a population variance (or standard deviation) are not built-in to TI calculators. These require critical values from a χ^2 (Chi-squared) distribution. Because the χ^2 distributions are not symmetric, these intervals are not of the usual *estimate* ± *ME*. Computation of these intervals is very sensitive to departures from the assumption the data comes from a normal distribution. This is a critical assumption to check – do a Normal quantile plot as described in Chapter 6.

EXAMPLE Home Voltage: The proper operation of home appliances requires voltage levels that do not vary much. Listed below are ten voltage levels (in volts) recorded at Dr. Triola's home on ten different days. These ten values have a standard deviation of s = 0.15 volt. Use the sample data to construct a 95% confidence interval estimate for σ, the standard deviation of all voltage levels (at his home).

123.3 123.5 123.7 123.4 123.6 123.5 123.5 123.4 123.6 123.8

With degrees of freedom (10 – 1) = 9, we can calculate the two critical values from the chi-square distribution which are required in this confidence interval calculation. Using a TI-83 or -84, we can use the Solver function from the MATH menu. The TI-89 and Nspire have an inverse Chi-squared function built in; use it just as in finding critical *t*-values.

Since χ^2 distributions are not symmetric, we need to find two critical values: χ^2_L with 0.025 area to the left, and χ^2_R with 0.975 area to the left. (The area between these points is our desired 95%.)

Finding χ^2 Critical Values TI-83/84 Procedure:

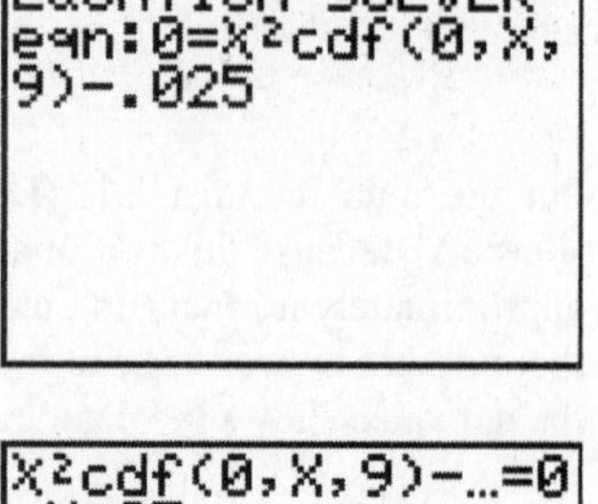

Press MATH then press the up arrow to find the Solver at option 0. Press ENTER to select the option. Press the up arrow to enter the equation to be solved.

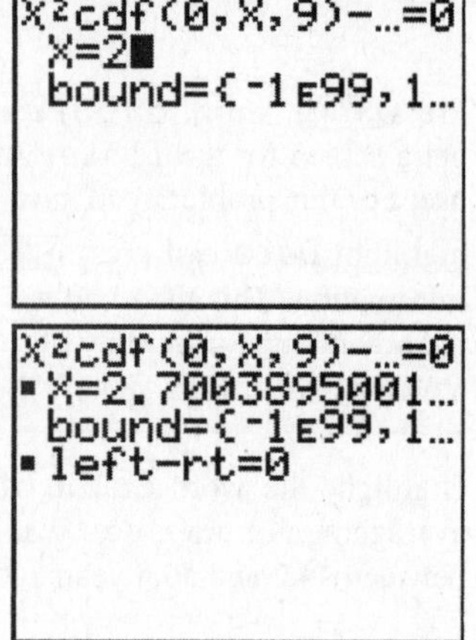

The equation as entered here will find χ^2_L, the point with .025 between 0 and X. Press ENTER to return to the Solver. You will need to enter a guess for X. This becomes easier when you know that the mean of a χ^2 distribution is its degrees of freedom. Since the mean for this particular distribution is 9, an appropriate guess is something much less (but positive, as these distributions cannot have negative values). I entered 2 as a guess.

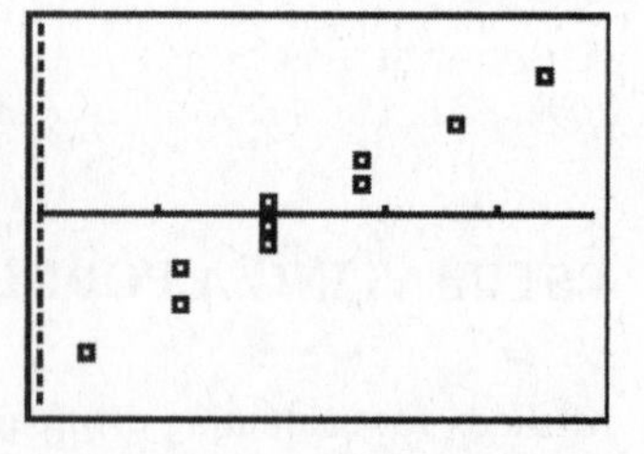

Press ALPHA ENTER to solve for X. (It may take a little while.) χ^2_L is 2.7003895. To find χ^2_R, repeat the procedure, but change 0.025 in the equation to 0.975 and use a larger guess (bigger than 9 – try 15, for example). I found χ^2_R to be 19.022767798.

Having found critical values, we should check the data to determine whether these may have come from a normal distribution. You could graph a histogram, but with only ten observations, this may not give a good picture. A better plot might be a normal quantile plot (see Chapter 6 for details). The graph at right does look like a stratight line, so a normal distribution for these data is reasonable. Using `1-Var Stats`, we find the standard deviation of these data is 0.149.

These results can now be used to calculate the confidence interval limits using the formula $\frac{(n-1)s^2}{\chi^2_R} < \sigma^2 < \frac{(n-1)s^2}{\chi^2_L}$. Our values are very close to the value obtained in the text, but more accurate as the text used *approximated* critical values. The variance of the voltage at Dr. Triola's home is between 0.0105 and 0.07399. Taking the square root of both ends, gives a 95% confidence interval for the standard deviation of the voltage as 0.102 to 0.272 volt.

```
9*.149²/19.02276
7798
      .0105036765
9*.149²/2.700389
5
      .0739926592
```

WHAT CAN GO WRONG?

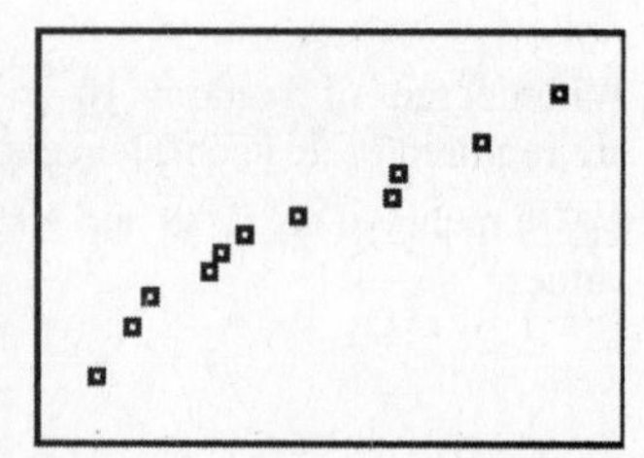

Assumptions not met

It is important to check that assumptions are met before conducting any inferential procedure. If for a proportion, are there at least 5 successes and 5 failures (so we can believe $\hat{p}$ is approximately normal)? In doing inference for a population mean, can we believe that $\bar{x}$ is approximately normal? You need a sample size of at least 30 for the Central Limit Theorem to apply. If you have a smaller sample, plot the data to check the assumption. This plot exhibits a skewed distribution. The inference methods we have discussed would not be appropriate.

Bad Interpretations

Be careful when interpreting the meaning of any confidence interval. One must include the level of confidence, the parameter of interest (what the question is about), and units of measure. An interval for a mean, for example, is the mean of the entire population of interest – not the mean of the sample (we *know* what that was), or an individual value.

8 Hypothesis Testing

In this chapter, you will learn to use your TI calculator to assist you in performing a variety of hypothesis tests based on one sample. You will see how to use the tests built into the STAT TESTS menu as well as how to calculate test statistics on the home screen. You will be able to handle samples in the form of raw data as well as summary statistics. When performing a test you should always make sure that all assumptions are met. Refer to your main text for this material. There are many options available, so make sure you are following a routine which will allow you to include all steps required by your instructor in your write-up of a test.

TESTING A CLAIM ABOUT A PROPORTION

EXAMPLE Gender Selection: ProCare Industries, Ltd. provided a product called "Gender Choice," which, according to advertising claims, allowed couples to "increase your chances of having a girl up to80%." Suppose we conduct an experiment with 100 couples who want to have baby girls, and they all follow the Gender Choice "easy-to-use in-home system" described in the pink package designed for girls. Assuming that Gender Choice has no effect and using only common sense to guide us, we deduce that we are to test the hypotheses: H_0: $p = 0.50$ and H_1: $p > 0.50$. Further suppose that these couples had 52 baby girls; that's more than 50%, but is it enough more to convince us that Gender Choice works?

TI-83/84 Procedure

We will use the built-in test for a population proportion from the STAT TESTS menu. This is option 5:1-PropZTest. This test yields output which is perfectly-suited for the p-value method of testing a hypothesis.

Press [STAT], arrow to TESTS, and select option [5]. Look at the inputs which are required. First is p_0, the value of the proportion *according to H_0*. This is 0.50 in our problem. Second is the value of x, the number in the sample with the characteristic of interest (here, a girl baby). We know that x =52 from the problem statement above. If given the sample results as a percentage, you can multiply to find x = $n\hat{p}$ and round, since there cannot be fractions of a success). You can enter the multiplication on the input screen, press [ENTER] to perform it, and then use the up arrow to do the necessary rounding. Third is the value of n which is 100. Fourth, we must choose the direction of our H_1. This is >pø in our problem. Choices correspond to two-tailed, left-tailed and right-tailed tests. Use the left or right arrows to move the cursor to the correct choice, and press [ENTER] to move the highlight. Finally, we have the option of simply calculating our test results or obtaining a drawing of our results. Either choice will do. Fill in the required inputs.

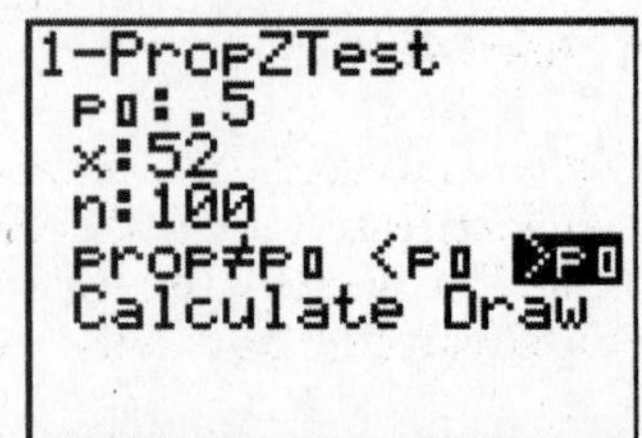

If you choose Calculate and then press [ENTER], you will see this screen. The computed *z*-statistic is 0.4. (Our observed 52% is less than 1 standard deviation above the claimed value of 50%. We know this is not a rare event). The p-value for the test (the probability of finding our observed 52% (or something more extreme), if the value really was 50% is 34.6%. This is large probability, and it leads us to believe the true percentage of girl babies for people using Gender Choice is probably not more than 50%; our observed 52% girls is likely to happen by sampling variation.

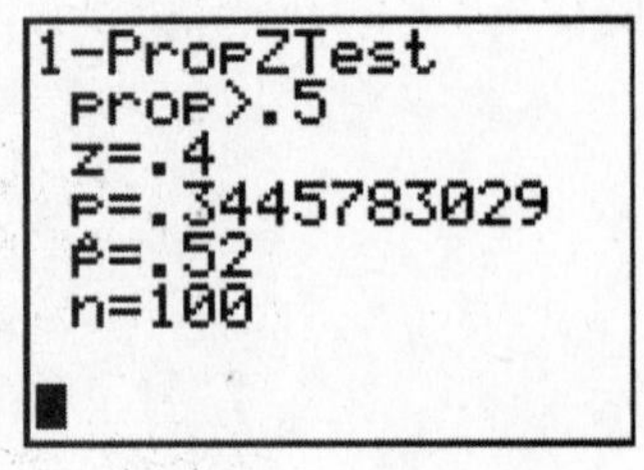

If you choose **Draw**, (first make sure that all **STAT PLOTS** are **OFF**) and then press [ENTER], you will see this screen. We are shown the z-statistic for the test and the p-value (to 4 significant digits). Notice the p-value is 0.3446. The area corresponding to the p-value of the test is shaded in the graph (like using **ShadeNorm** for normal calculations).

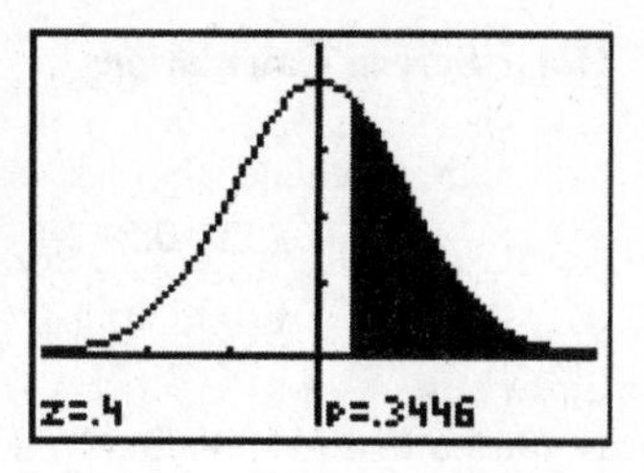

Our conclusion about Gender Choice? We will conclude that it most likely does *not* work as claimed in helping couples have girl babies, based on this set of data.

TI-89 Procedure

Input the values from our data. If you need to multiply to find x = $n\hat{p}$, this must be done on the home screen.

From the **Statistics/List Editor**, press [2nd][F1] (= [F6] **Tests**). Press [5] to select **1-PropZTest**. First is p0, the value of the proportion *according to H_0*. This is 0.50 in our problem. Second is the value of x, the number sampled with the characteristic of interest. That was given as 52. Third is the value of n which is 100. Fourth, we must choose the direction of our H_1. Use the right arrow key to expand and select the proper direction. Lastly, you can decide to simply calculate the results, or display a normal curve with the area corresponding to the p-value shaded.

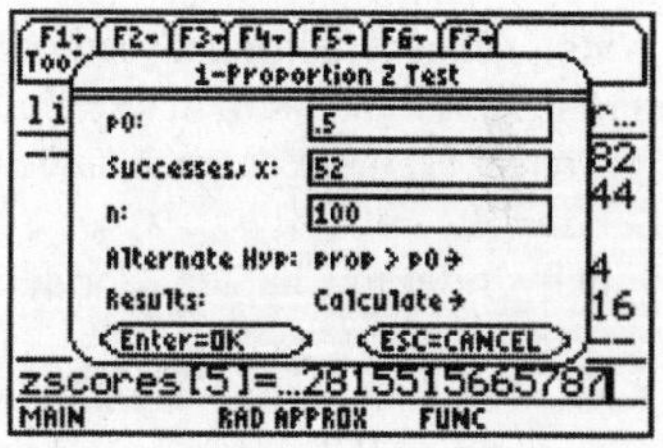

The screen at right is the output from selecting **Calculate**. The results are identical to those obtained with an 83 or 84.

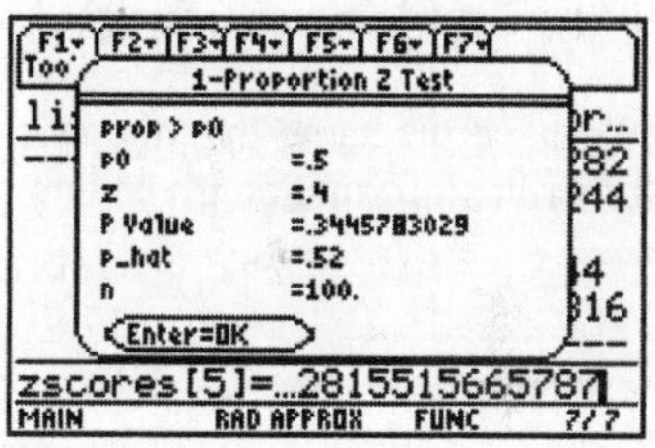

TI-Nspire Procedure

All these procedures are similar on all models of TI calculators. We illustrate this example on the Nspire. Note that as with intervals, you can find the statistical tests either from the spreadsheet or calculations applications windows. If done from the spreadsheet, the results will be placed in the first blank column of the spreadsheet; you can select to draw the curve to illustrate the p-value by clicking on the Plot data button to put a check in the box.

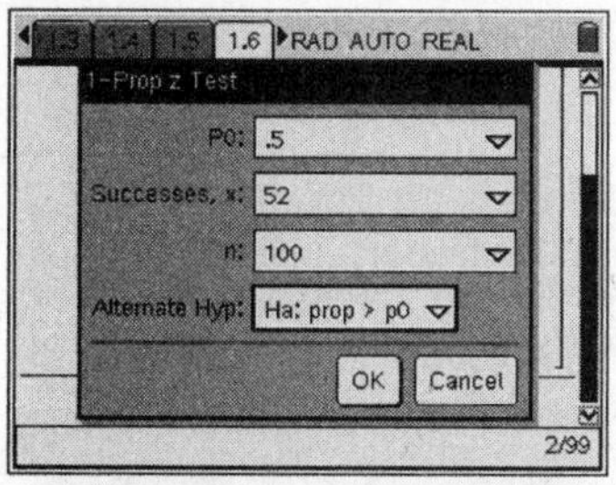

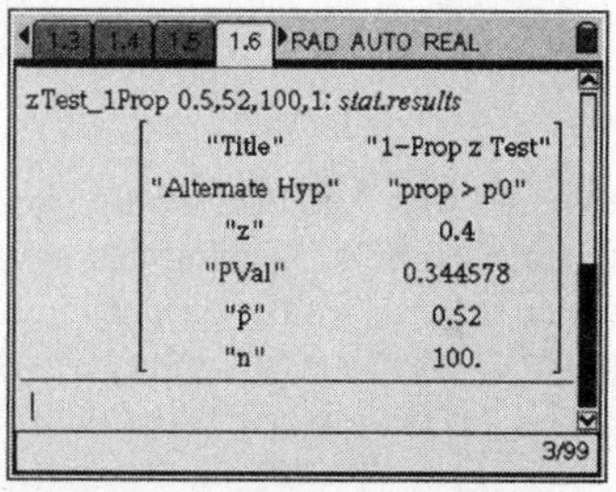

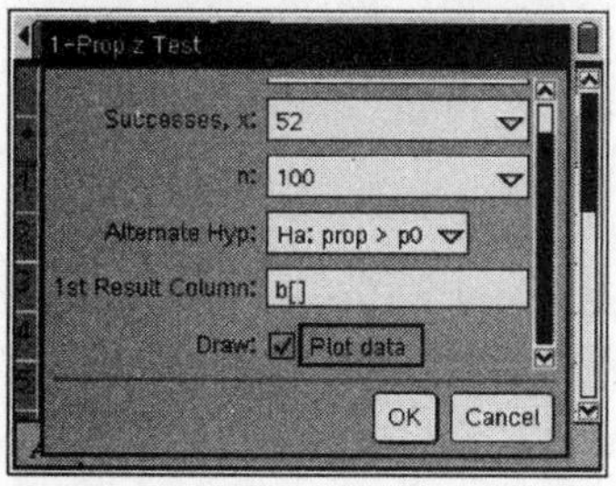

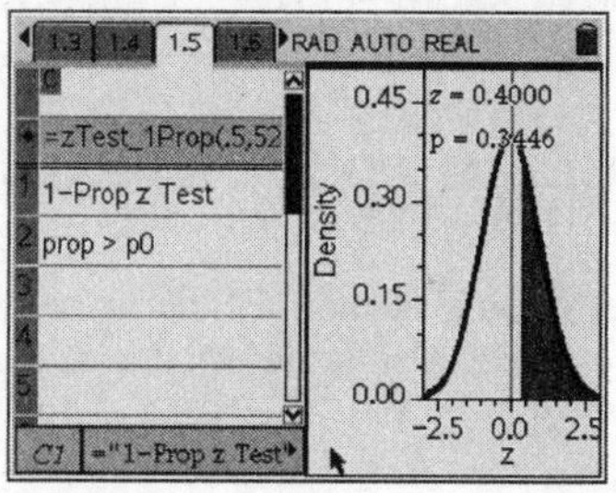

Home Screen Calculations

You can calculate the test statistic on the home screen using the formula $z = \dfrac{\hat{p} - p_0}{\sqrt{\dfrac{p_0 q_0}{n}}} = \dfrac{0.52 - 0.5}{\sqrt{\dfrac{(0.5)(0.5)}{100}}}$. We find our test statistic is $z = 0.4$. (Any difference would be due to rounding You can calculate the p-value of this test statistic by finding the area (probability) in the tail of the standard normal distribution corresponding to the direction of the alternate hypothesis (here, greater than the observed value) past the statistic as in Chapter 6.

```
(.52-.5)/√(.5*.5
/100)
                .4
normalcdf(.4,E99
,0,1)
      .3445783029
```

Also (for those using the Traditional Method) you can calculate the critical value associated with a particular significance level (in this case $\alpha = 0.05$) using the command `invNorm(.95`, since on a right-tailed test like this, our α area is on the right side of the curve, so 0.95 is the area to the *left* of the critical value.

```
invNorm(.95
      1.644853626
```

Note: The Traditional Method would lead us to compare the critical value 1.645 and the test statistic 0.4. We find that the test statistic is less and thus leads us to *not* reject H_0 in favor of H_1 since this is a right-tailed (>) test.

Finally, refer to Chapter 7 for details if you wish to use the Confidence Interval Approach to testing.

EXAMPLE Another Gender Selection Method: The XSORT method of gender selection developed by the Genetics and IVF Institute had 668 girl babies born to 726 couples wanting a girl. Is this evidence that this method of gender selection is effective?

We have the same hypotheses as before, namely, H_0: $p = 0.50$ and H_1: $p > 0.50$. Enter the information given into the `1-PropZtest`.

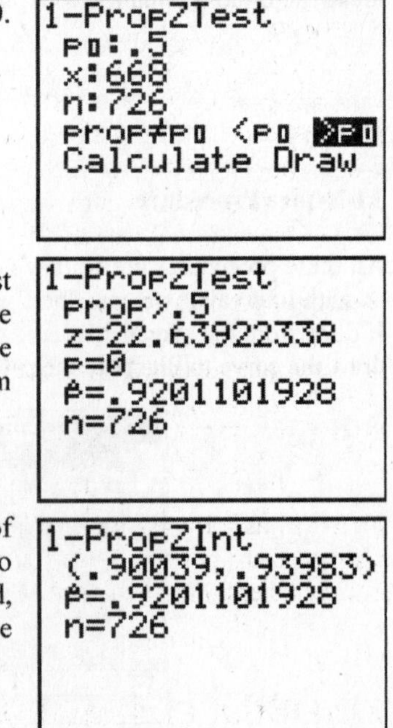

We find in this case, that 668/726 = 92% of these babies were girls. Our test statistic is $z = 22.64$ and the p-value of the test is 0. In this case, we have overwhelming evidence that this method of gender selection is effective; if chance alone were at work, we'd practically never expect to see results this far away from $p = 0.5$.

Since we are convinced that XSORT works, we might wonder what percent of babies could be expected to be girls using this method. If we use the data given to find a 95% confidence interval for the proportion of girls with the XSORT method, we find this is 90.0% to 94.0%. This interval (well above 50%) is also evidence that the likelihood of a girl baby is increased with this method of gender selection.

TESTING A CLAIM ABOUT A MEAN, σ KNOWN

EXAMPLE Overloading Boats: People have died in boat accidents because an obsolete estimate of the mean weight of men was used. Using the weights of the simple random sample of men from Data Set 1 in Appendix B, we obtain these sample statistics: $n = 40$ and $\bar{x} = 172.55$ lb. Research from several other sources suggests that the population of weights of men has a standard deviation given by $\sigma = 26$ lb. Use these results to test the claim that men have a mean weight greater than 166.3 lb, which was the mean weight in the National Transportation and Safety Board's recommendations M-04-04. Use the 0.05 significance level.

As before when you worked with confidence intervals for a mean, you have two options for input. You can input summary statistics or you can use the raw data as input. If you wish to use raw data, you must have it stored in a list.

Press STAT ▶ ▶ 1 to get the input screen (2nd F1 = F6 (Tests) followed by ENTER on an 89). Enter the sample statistics and press ENTER on Calculate.

In this TI-89 dialog box, we have chosen to input summary statistics. The inputs are similar to those on the TI-83/-84. Fill in the rest of the information requested. Note this is a one-tailed test (the claim is that people now weigh more, on average), so we choose > μ_0 as our alternative hypothesis.

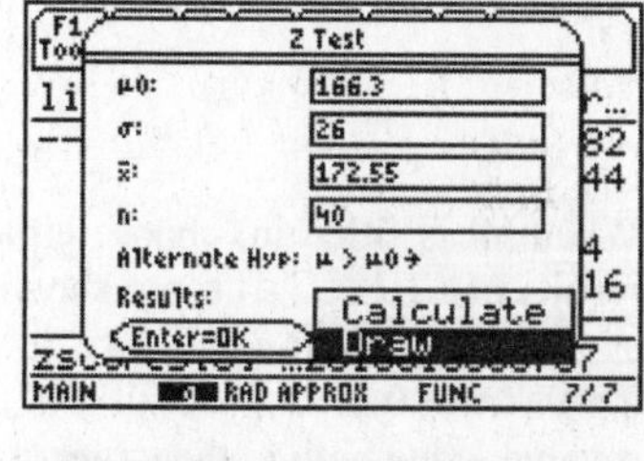

When all is filled in, choose either Calculate or Draw and press ENTER. The results of "Draw" (this author's personal favorite option) are shown. Our computed z-statistic is $z = 1.52$, with a p-value of 0.06422. This probability is larger than our stated level of significance; it tells us that there is a moderately good chance of obtaining our observed sample mean ($\bar{x} = 172.55$ lb) if the population mean is 166.3 lb. We fail to reject the null hypothesis; we are not convinced the average man weighs more now than when the National Transportation and Safety Board's recommendations M-04-04 was written. You may have slightly different values (due to rounding on the summary statistics) but the conclusion is still the same.

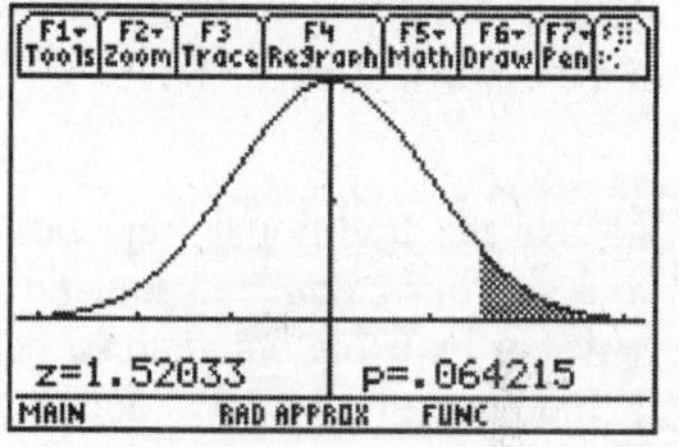

Home Screen Calculation

We have reproduced the calculations for the home screen calculations for the test statistic $z = \dfrac{\bar{x} - \mu_0}{\dfrac{\sigma}{\sqrt{n}}}$ and its p-value at right. The author does not recommend this approach, as misplacing parentheses and failing to double the p-value for a two-sided test are potentially fatal errors — I recommend the automated approach. Among other things, a misplaced parenthesis can lead to drastically wrong results.

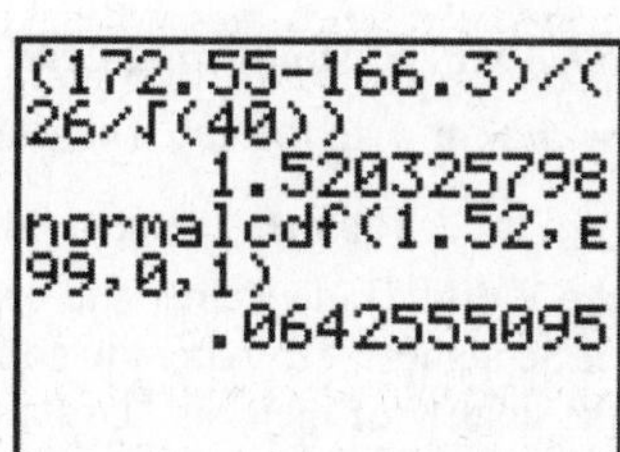

TESTING A CLAIM ABOUT A MEAN, σ NOT KNOWN

EXAMPLE Overloading Boats: In the example in the previous section, we assumed a population standard deviation σ, based on other studies. Now, use the standard deviation of the sample in Appendix B (s = 26.33 lb) to recompute the test.

We will use the calculators' built-in test for a population mean when σ is not known. This is option 2 from the **STAT TESTS** menu. It is the `T-Test`. As in the previous examples, this test yields output which is suited for the p-value method of testing a hypothesis.

As always, you have two options for input. You can input summary statistics (as we have here), or you can use the raw data as input. If you wish to use raw data, you must have it stored in a list. For this example, we have summary statistics.

Press STAT ▸ ▸ 2 (option 2 on the **TESTS** menu on an 89 or Nspire). Here, I have chosen `Stats` as the input method. To move the highlight, move the blinking cursor to your choice and press ENTER.

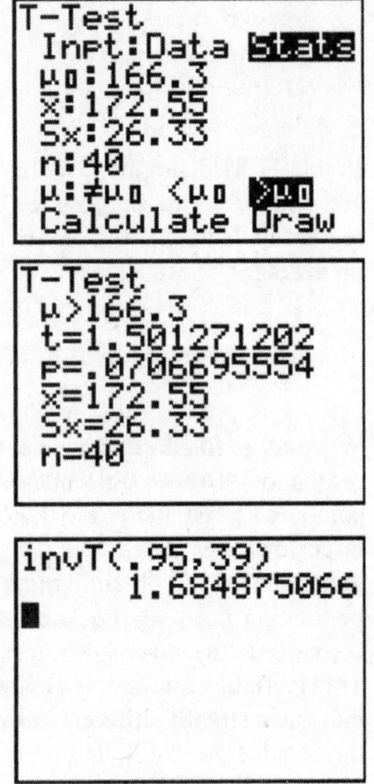

Specify the *hypothesized mean* $\mu_0 = 166.3$. This time we do not have to specify the population standard deviation σ (but we do have to input *s* because we have chosen the `Stats` input method. Note this is a right-tailed test, so we choose $>\mu_0$ as our alternative hypothesis.

When all is filled in, choose either `Calculate` or `Draw` and press ENTER. The results of `Calculate` are shown at right. Our test statistic is $t = 1.50$ (slightly smaller than that found before due to the somewhat larger standard deviation), with a p-value of 0.0707. We still fail to reject the null hypothesis (with a slightly larger p-value – due both to the larger standard deviation and *t* distribution), and conclude that the mean weight of men has not significantly increased.

To use the traditional (rejection region) approach, we need a *t*-critical value to compare our calculated statistic against. The critical value can be obtained from tables or by using equation solver as described in Chapter 7 (TI-83) or the Inverse T function (TI-84, -89, and Nspire). In the screen at right, I have done the computation using a TI-84. The parameters are the area to the left of the desired point (since our test at $\alpha = 0.05$ was right tailed, all our α area is on the right, so 0.95 to the left) and with $n = 40$ pieces of data, we have $n - 1 = 39$ degrees of freedom. Since our computed statistic ($t = 1.50$) is less than the critical value, we fail to reject the null hypothesis.

EXAMPLE Monitoring Lead in Air: Listed below are measured amounts of lead (in micrograms per cubic meter, or $\mu g/m^3$) in the air. The Environmental Protection Agency has established an air quality standard for lead of 1.5 $\mu g/m^3$. The measurements shown constitute a simple random sample of measurements recorded at Building 5 of the World Trade Center site on different days immediately following the destruction caused by the terrorist attacks of September 11. After the collapse of the two World Trade Center buildings, there was considerable concern about the quality of their air. Use a 0.05 significance level to test the claim that the sample is from a population with a mean greater than the EPA standard of 1.5 $\mu g/m^3$.

5.40 1.10 0.42 0.73 0.48 1.10

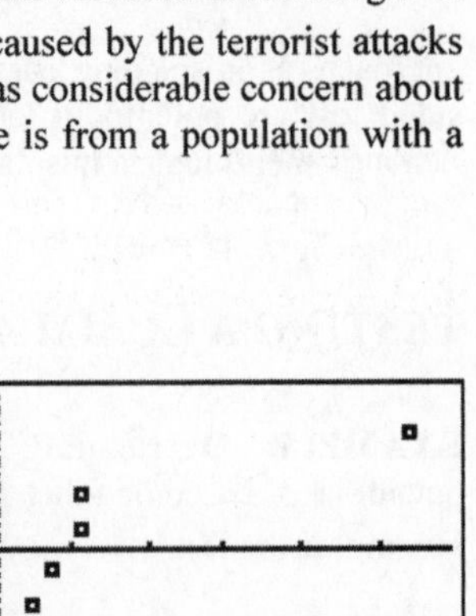

This is a small sample; just looking at the data it would seem that the first value (5.40) is possibly an outlier, since it is much larger than the rest. With this small sample, we can create a normal quantile plot to assess whether we believe these data came from a normal population. See Chapter 6 for full details. We can clearly see the gap and bend in the plot; these data are not suited for a *t*-test of hypotheses. This is an example that points out the fact that checking assumptions is critical!

TESTING A CLAIM ABOUT A STANDARD DEVIATION OR VARIANCE

Just as with confidence intervals for a variance or standard deviation, there is no built-in function on TI-calculators to conduct a hypothesis test for a variance or standard deviation. The Chi-square cdf function on all calculators can help find exact p-values, however, and the Equation Solver (on TI-83/84s) and Inverse Chi-Square on TI-89s can help find critical values. As with confidence intervals for a variance, this procedure is very sensitive to any departures from the normal distribution assumption – be sure to check plots of your data to be sure this is satisfied!

EXAMPLE Quality Control of Coins: The industrial world shares this common goal: Improve quality by reducing variability. Quality-control engineers want to ensure that a product has an acceptable mean, but they also want to produce items of consistent quality so that there will be few defects. If weights of coins have a specified mean but too much variation, some will have weights that are too low or too high, so that vending machines will not work correctly. Consider the simple random sample of 37 weights of post-1983 pennies listed in Data Set 20 in Appendix B. Those 37 weights have a mean of 2.49910 g and a standard deviation of 0.01648 g. A hypothesis test will verify that the sample appears to come from a population with a mean of 2.500 g as required, but use a 0.05 significance level to test the claim that the population of weights has a standard deviation less than the specification of 0.0230 g.

```
36*.01648²/.0230
²
         18.4825225
χ²cdf(0,18.4825,
36)
       .0068732107
```

Calculate the test statistic from the following equation:

$\chi^2 = (n-1)\dfrac{s^2}{\sigma^2} = (37-1)\dfrac{.01648^2}{.0230^2} = 18.4825$ as seen in the top of the screen.

The bottom of this screen shows the calculation of the p-value of the test statistic using the χ^2cdf function from the `DISTR` menu. Note we wanted area *below* our calculated statistic because the claim is that the standard deviation is now *lower*.

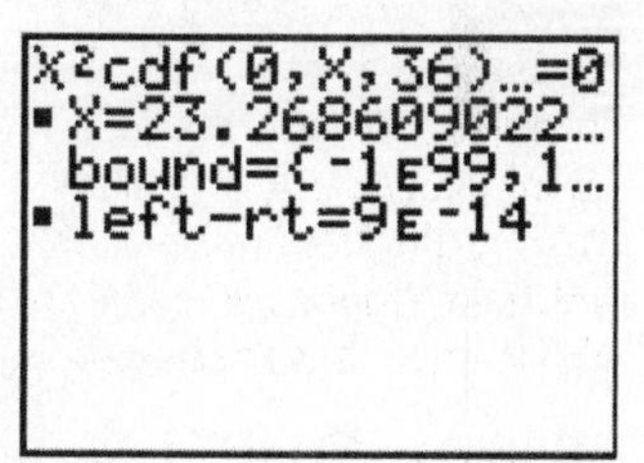

Finally, for those using the traditional method, you can find the critical values using the Equation Solver (TI-83/84) or Inverse Chi-Square as detailed in Chapter 7.

The p-value is 0.0069 which is less than the significance level of the test. The test statistic (18.4825) is less than the left critical value of the test (23.2686), so either way you look at it, the sample has provided evidence that the standard deviation of the population from which these coins were sampled is less than the specification of 0.0230 g.

WHAT CAN GO WRONG?

Err: Domain?

This error stems from one of two types of problems. Either a proportion was entered in a `1-PropZTest` which was not in decimal form or the numbers of trials and/or successes was not an integer. Go back to the input screen and correct the problem.

```
ERR:DOMAIN
1:Quit
```

Err:Invalid Dim?

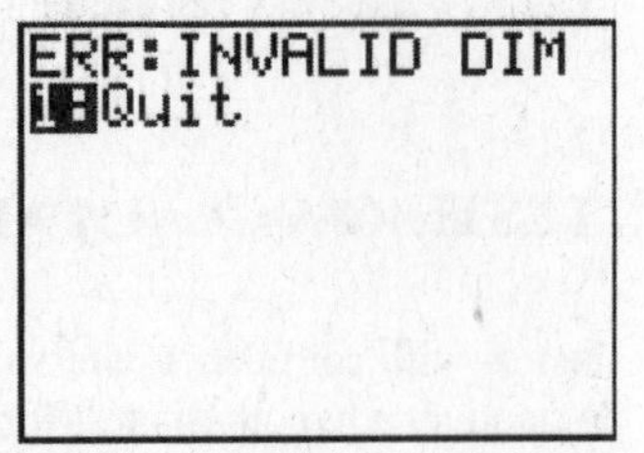

This can be caused by selecting the DRAW option if another Statistics plot is turned on. Either go to the STAT PLOT menu (2nd Y=) and turn off the plot or redo the test selecting CALCULATE.

Bad Decisions.

It is important to remember that we have calculated a test statistic *assuming the null hypothesis is true*. The p-value of the test is the probability of our sample result (or something more extreme) under that assumption. If our sample result is unlikely to have happened, we reject the null hypothesis in favor of the alternate (because we have faith our sample represents the "truth" of the situation). That is, reject H_0 for ***small*** p-values.

Bad Conclusions.

If the p-value is large, the null hypothesis is not rejected; but this does *not* mean it is true – we simply haven't gotten enough evidence to show it's wrong. Be careful when writing conclusions to make them agree with the decision – remember, the null hypothesis says the claimed value is *true*. If H_0 is rejected, this means you no longer believe this value, but rather believe the true value (mean, proportion, or variance) is different (or higher, or less).

9 Inferences from Two Samples

In Chapters 7 and 8 you found confidence intervals and tested hypotheses for data sets that involved only one sample from one population. In this chapter you will learn how to extend the same concepts for use on data sets that involve two samples from two populations. As in Chapters 7 and 8 you will use functions from the `STAT TESTS` (or `INTS`) menu. The presentation in this chapter assumes a familiarity with the materials presented in Chapters 7 and 8. As always, if there are significant differences, the TI-89 (or Nspire) procedure is explained immediately after that for the TI-83 and -84.

INFERENCES ABOUT TWO PROPORTIONS

EXAMPLE Do Airbags Save Lives?: The table below lists results form a simple random sample of front-seat occupants involved in car crashes (based on data from "Who Wants Airbags?"). Use a 0.05 significance level to test the claim that the fatality rate of occupants is lower for those in cars equipped with airbags. We notice both sample proportions are small; is this small difference (less than 0.2%) statistically meaningful?

	Airbag	No airbag
Occupant Fatalitites	41	52
Total number of occupants	11,541	9,853
Fatality Rate	0.36%	0.53%

Hypothesis Test: 2-PropZTest

Press STAT ▶ ▶ 6 to get to the `STAT TESTS` menu and choose option 6, the `2-PropZTest` (on a TI-89, from the Statistics/List Editor press 2nd F1 6). The first four input values come directly from the table. Next we choose the direction of our test hypothesis. We are asked to test if the proportion of fatalities with airbags is *less* than without airbags. Since we represented the surgery patients with `x1` and `n1`, it would follow that our test hypothesis is p1<p2. Thus you should fill in the screen as at right. Input screens for the TI-89 and Nspire are similar.

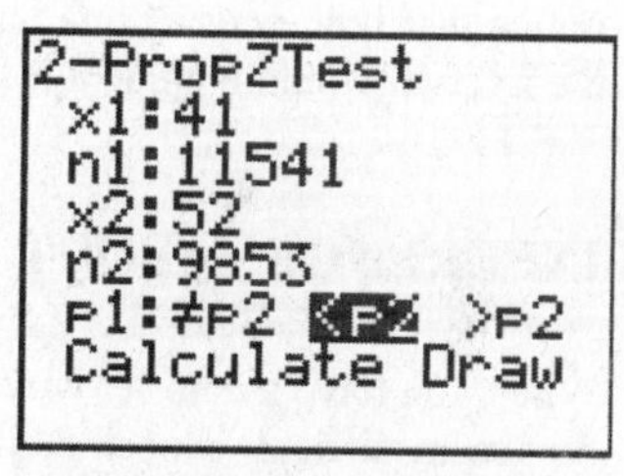

Highlight the word `Calculate` and press ENTER for the results. The test statistic is $z = -1.91$. Note the p-value is 0.0279. Since this p-value is less than the 5% significance level, we will reject the null hypothesis and conclude that airbags do reduce the risk of fatalities. Although this difference might be small (only about 0.2%, preventing deaths is important.)

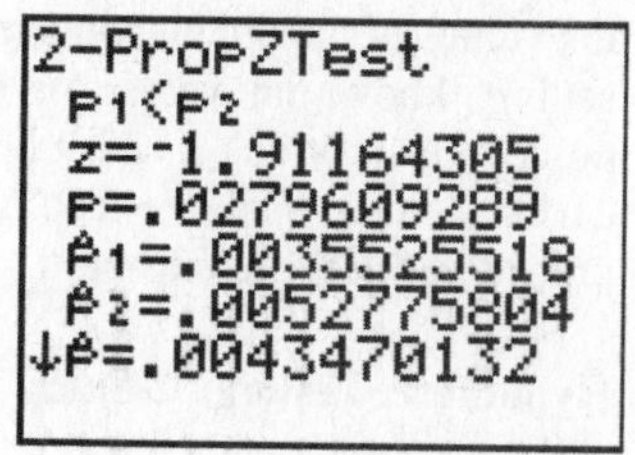

If one really wanted to, one could also do the calculation of the test statistic on the home screen. This first involves the calculation of a "blended" proportion (because if the two are really the same, we could combine our results into a single sample) of $\bar{p} = \frac{x_1 + x_2}{n_1 + n_2} = \frac{41+52}{11541+9853} = 0.0043$. This was shown as $\hat{p}$ at the bottom or our results screen. Then we find the test statistic

$$Z = \frac{p_1 - p_2}{\sqrt{\bar{p}(1-\bar{p})\left(\frac{1}{n_1} + \frac{1}{n_2}\right)}}$$. Again, we obtain $z = -1.92$ (to within rounding error).

Since the p-value of the test is less than the significance level of 0.05, we find that our data has provided us with significant evidence that airbags do help prevent fatalities in car crashes.

Confidence Interval: 2-PropZInterval

EXAMPLE Confidence Interval for Airbags: Use the sample data in the example above to construct a 90% confidence interval for the difference in the two population proportions. (A 90% confidence interval is similar to a 95% 1-tailed hypothesis test.)

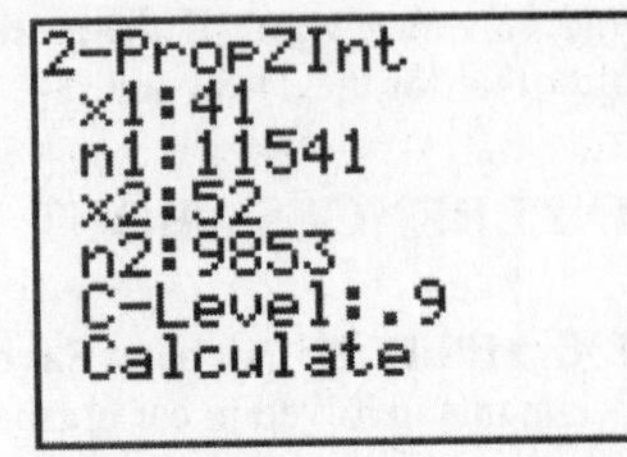

Press [STAT] [▶] [▶] for the STAT TESTS menu. Choose Option B which is the 2-PropZInt. (This is option 6 on the Ints menu on a TI-89 or Nspire.) Fill in the screen as shown.

Highlight Calculate and press [ENTER] for the interval. We obtain the results (–0.0032, –0.0002). To retrieve the margin of error we can simply find the difference in the endpoints and divide by 2. $E = (0.0002 - (-0.0032))/2 = 0.0017$.

We find that somewhere between 0.02% and 0.32% fewer fatalities with airbags, at 90% confidence. Be sure to notice that both ends of this interval are negative. If there were no difference in fatalities with or without airbags, the interval would include 0, which is not the case here.

INFERENCES ABOUT TWO MEANS: INDEPENDENT SAMPLES

When you have a data set which is comprised of values from two independent samples of two populations it can at first seem difficult choosing how to begin. For the problems in your Triola text it is assumed that you do not know the values of the population variances σ_1^2 and σ_2^2. (In real life, this is almost always the case.) If you, in some future setting, know the values of the population variances you can use the 2SampZTest and the 2SampZInt procedures (options 3 and 9 on the TI-83/84 STAT TESTS menu, option 3 on the TI-89 and Nspire Tests and Ints menus). With the idea of keeping things simple, we will not go into more detail or show examples on these procedures in this companion.

Hypothesis Testing: 2-SampTTest (Assuming $\sigma_1 \neq \sigma_2$, σ_1 and σ_2 not known)

EXAMPLE Are Men and Women Equal Talkers?: A headline in *USA Today* proclaimed "Men, women are equal talkers." That headline referred to a study of the numbers of works that samples of men and women spoke in a day. Given below are the results from the study, which are included in Data Set 8 in Appendix B (based on "Are Women Really More Talkative than Men?" by Mehl, et al, *Science*, Vol. 317, No. 5834). Use a 0.05 significance level to test the claim that men and women speak the same number of works in a day (on average). Does there appear to be a difference?

Number of Words Spoken in a Day	
Men	Women
$n_1 = 186$	$n_2 = 210$
$\bar{x}_1 = 15{,}668.5$	$\bar{x}_2 = 16{,}215.0$
$s_1 = 8632.5$	$s_2 = 7301.2$

We use the `2-SampTTest` (option 4) from the `STAT TESTS` menu to perform the test. We have summary statistics; enter them after selecting the `Stats` input mode.

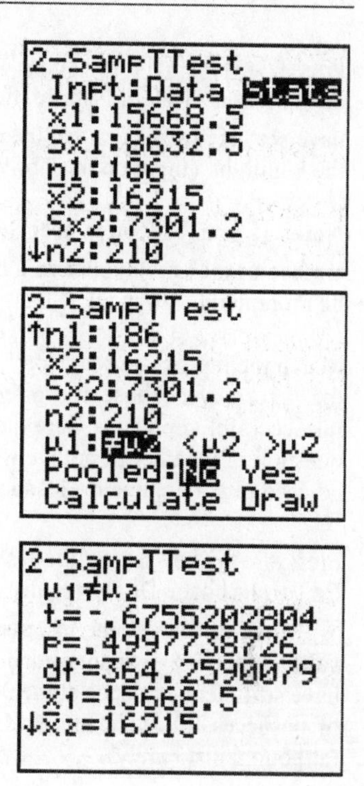

We also select the appropriate alternate hypothesis ($\neq$ since we want to know if there is a difference) and select No for the pooled question (we do not assume the two *populations* have the same standard deviation).

Press [ENTER] on `Calculate`. We find the test statistic is $t = -0.68$ with a p-value for the test of 0.4998. This large p-value is an indication that our observed difference in mean words spoken per day between men and women is likely to occur by chance. We fail to reject H_0, and conclude that men and women do, on average, talk the same amount per day.

EXAMPLE Discrimination Based on Age: The revenue commissioners in Ireland conducted a contest for promotion. The ages of unsuccessful and successful applicants are given below (based on data from Barry and Boland, *The American Statistician*, Vol. 58, No. 2). Some of the applicants who were unsuccessful in getting the promotion charged that the competition involved discrimination based on age. If we treat the data as samples from larger populations and use a 0.05 significance level to test the claim that the unsuccessful applicants are from a population with a *greater* mean age than the mean age of unsuccessful applicants, does there appear to be discrimination based on age?

Ages of Unsuccessful Applicants									
34	37	37	38	41	42	43	44	44	45
45	45	46	48	49	53	53	54	54	55
56	57	60							

Ages of Successful Applicants									
27	33	36	37	38	38	39	42	42	43
43	44	44	44	45	45	45	45	46	48
47	47	48	48	49	49	51	51	52	54

I have entered the ages of the unsuccessful applicants into list `L1` and the ages of the successful applicants into list `L2`. The first step with these small samples is to determine whether they each come from approximately normal populations. Create a normal quantile plot (see Chapter 6 for details) for each set of data. Both plots are reasonable close to straight lines, so we may proceed.

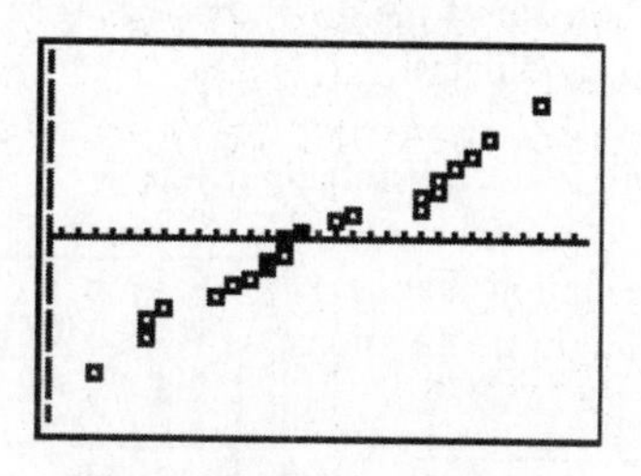
Unsuccessful Applicants

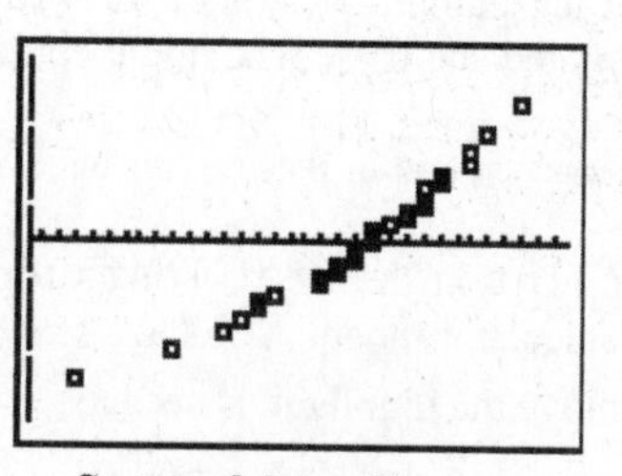
Successful Applicants

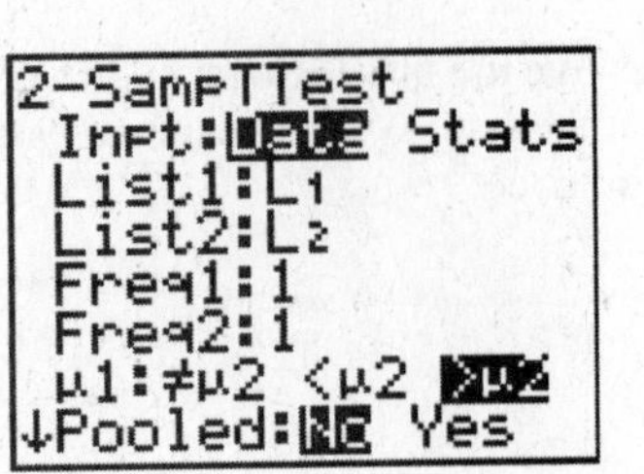

Press [STAT] [▶] [▶] [4] to choose Option `4:2-SampTTest` from the `STAT TESTS` menu. (On a TI-89 or Nsprire, this is also option 4 on the `TESTS` menu). Since we have sets of raw data, move the cursor to highlight `Data` and press [ENTER] to move the highlight. (Input on the TI-89 is similar.)

Note: If you had only been given summary statistics (as is the case in some textbook problems), you would highlight `Stats` and press [ENTER]. Then you would be prompted to fill in all of the statistics for both data sets.

Match the inputs you see. Note we are telling the calculator which lists contain our two data sets. Near the bottom of the screen we choose the direction of the test hypothesis. This is a *one-tailed* test because the claim is that unsuccessful applicants were older. Also in the very bottom of the screen, we have chosen the answer "No" to the question of whether or not we will *pool* our standard deviations. (For homework exercises 9-28 in your text, you are advised to not assume the standard deviations are equal. This equates to not pooling.).

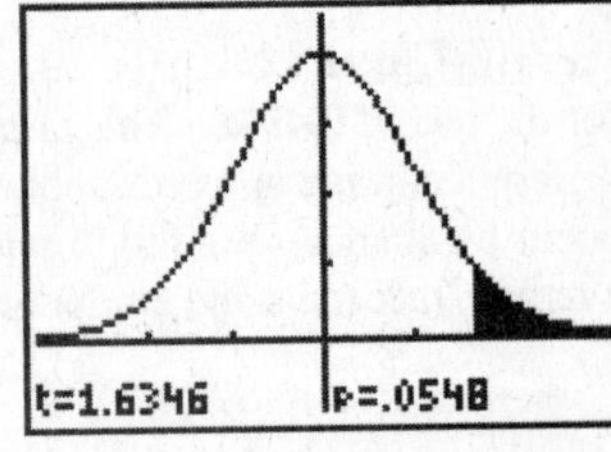

Press the [▼] key to see the last option. As usual for tests, you can choose either `Calculate` or `Draw` for displaying the results. We chose "Draw" and pressed [ENTER]. You can see the test statistic in the lower left corner is $t = 1.6346$. The p-value of 0.0548 is displayed in the lower right corner. It follows that we do not have sufficient evidence at the 0.05 significance level to conclude that the mean age of unsuccessful applicants was older than the mean age of successful applicants (although, just barely).

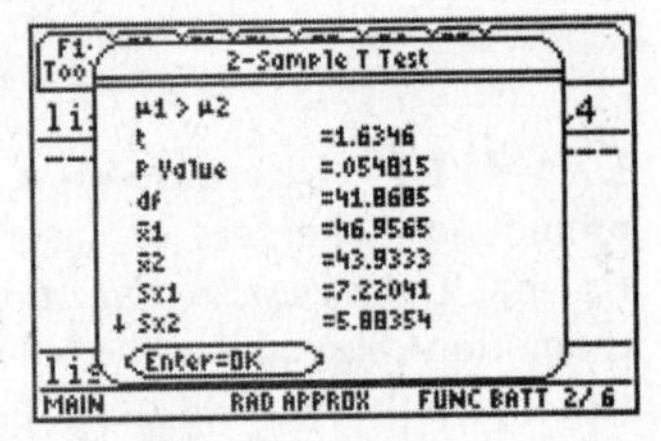

Your text suggests using the conservative approach to p-values for this situation. TI calculators (and most software) use a complicated formula for the degrees of freedom for these tests. As we can see in the screen at right (having chosen the `Calculate` option), the degrees of freedom (41.8685) are not even an integer.

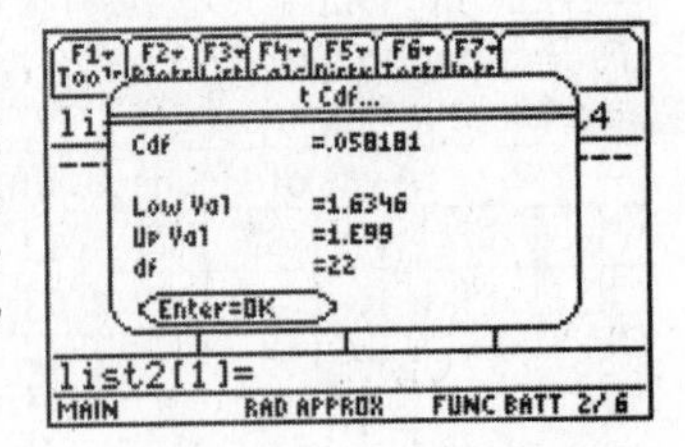

To find a p-value using the conservative approach (the smaller of $n_1 - 1$ and $n_2 - 1$ which would be 22 in this example) we can use the `Tcdf` function from the `DISTR` menu. The results are displayed at right. This p-value does not change the conclusion; however there are cases in which this might occur. *Be sure to double the p-value obtained in this manner if you are working with a two-tailed alternate hypothesis.*

Confidence Interval: 2-SampTInterval (Assuming $\sigma_1 \neq \sigma_2$)

EXAMPLE Confidence Interval for Ages of Applicants for Promotion: Again using the data sets from on age of applicants for promotion, construct a 90% confidence interval estimate of the difference between the mean ages of those not promoted and those who might be successful promotion candidates.

The two data sets are stored in lists `L1`(unsuccessful applicants) and `L2` (successful applicants).

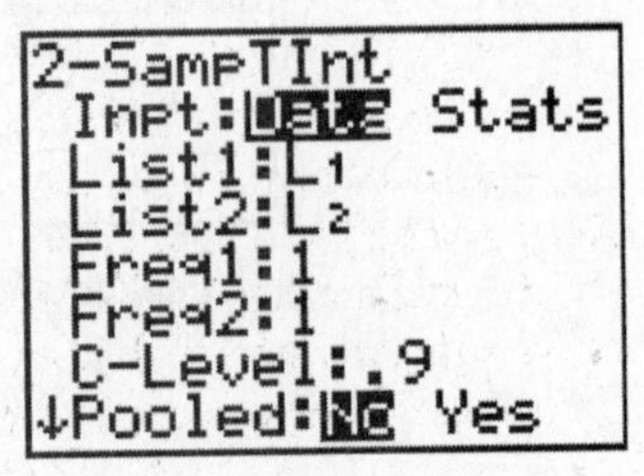

Locate `2-SampTInt` in the `Stat Tests` (or `INTS`) menu; the option number varies with the calculator model. You have sets of raw data, so highlight `Data` and press [ENTER] to move the highlight, if needed.

Note: If you had only been given summary statistics (as is the case in some textbook problems), you would highlight `Stats` and press [ENTER]. Then you would be prompted to fill in all of the statistics for both data sets. Fill in the rest of the information to duplicate what you see in the screen. Note that near the bottom we indicate that our desired confidence level is 90%. Again, we answer No to the pooling question.

Press the ▽ key to highlight Calculate and press ENTER. You can see the confidence interval is (-.0878, 6.1341). We are 90% confident that the difference in the mean ages is between -.0878 and 6.1341 years. Since 0 is included in the interval, there is not a significant difference in mean age between those not promoted and those promoted.

```
2-SampTInt
 (-.0878,6.1341)
 df=41.86847492
 x̄1=46.95652174
 x̄2=43.93333333
 Sx1=7.22041462
↓Sx2=5.88354417
```

Home Screen Calculations – Conservative Approach.

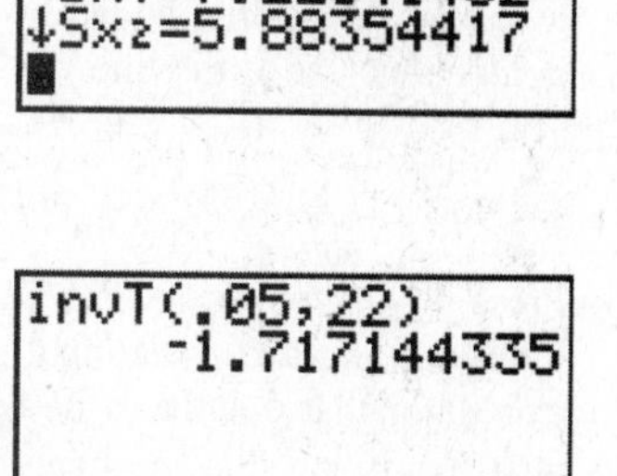

The confidence interval computed with the built-in function uses the computed degrees of freedom. If you want to use the conservative approach, you will have to compute the interval "manually." You will first need a critical t-value – one that puts 5% (for 90% confidence, $\alpha = .10$, so $\alpha/2 = .05$) in each tail of the distribution. The smaller sample was for those not promoted, $n = 23$, so df = 22. If you have a TI-84 or TI-89, you can use the built-in invT function to find the critical value. If you have a TI-83, either use a table or the Equation Solver as detailed in Chapter 7 to find the critical value.

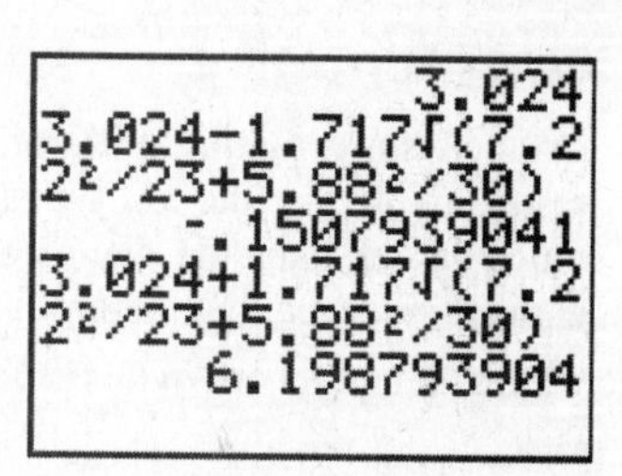

We first found the difference in the two sample means (46.957 – 43.933 = 3.024, using the means in the screen above). The endpoints of the interval are then calculated using the formula $\bar{x} \pm t_{\alpha/2}\sqrt{\frac{s_1^2}{n_1}+\frac{s_2^2}{n_2}}$.

From the screen at right, we see the conservative 90% confidence interval for the difference in mean age is (-.15, 6.20). As always, slight differences are due to rounding.

INFERENCES ABOUT TWO MEANS: MATCHED PAIRS (DEPENDENT SAMPLES)

EXAMPLE Is the Freshman 15 Real?: The table below consists of actual April and September weights (in kilograms) for the same students. The data consists of matched pairs, because each pair of values represents the same student. Here, we want to know if there is sufficient evidence to conclude that there is a weight change for students in their freshmen year in college. Use a 0.05 significance level to test the claim that there is a *difference* between the April and September weights.

April Weight	66	52	68	69	71
September Weight	67	53	64	71	70
Difference (April – September)	–1	–1	4	–2	1

We see that the above problem leads to the following hypotheses: H_0: $\mu_d = 0$ (no weight difference on average) versus H_1: $\mu_d \neq 0$ (there is a difference in weight).

Put the April weights in L1 and the September weights in L2. Store the differences (L1-L2) in L3. Do this by highlighting L3 in the Statistics Editor and then typing L1-L2 and pressing ENTER. The results should be the same as those seen in the bottom row of the above table. (Actually, since the differences were given, you could have entered these directly, but we are illustrating a general method for this type of test.) Now you are ready to treat the differences as a single data set on which you can perform the desired test. The procedure for obtaining the differences is the same on the TI-89. If you are using an Nspire, enter the formula in the compute line of the spreadsheet as at right. This can be done by referring to either the variable names (as in apr-sep) or using the column letter names as shown.

This is a small sample again. We (as always) need to check whether or not the assumption that the data (here, the data are the differences) came from a normal population is satisfied. A normal quantile plot is fairly straight, so we may continue. (Actually, this plot does show some curvature, but with the very small sample size, this is OK, as long as there are no clear outliers, and that is the case here. With the line added by the Nspire for reference, it is clear that all data points are close to the line.)

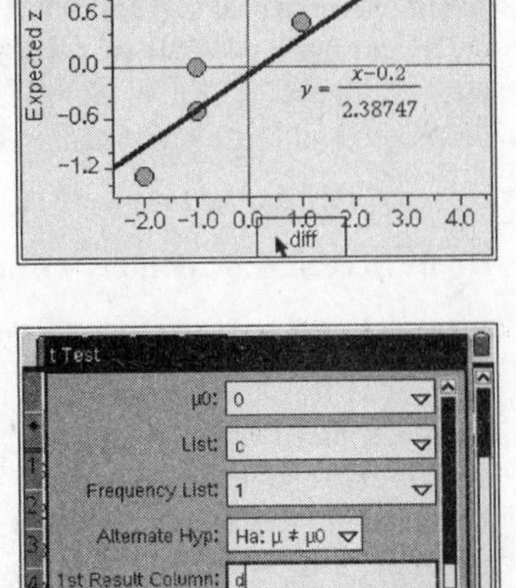

Press STAT, arrow to TESTS, and select option 2:T-Test; use the Data input option. Set up the input screen to match mine. Note that we are telling the calculator that our data (the differences) is in column c (or list L3). We are also specifying that the hypothesized mean μ_0 is 0 and that this is a two-tailed test (we want to know if the mean difference is *not* 0).

On the Nspire, highlight OK (or Highlight Calculate) and press ENTER for the results. Note that the test statistic is $t = 0.187$ and the p-value is 0.8605. This is much larger than the significance level of 0.05, so our data was not strong enough evidence that there was an actual difference in weight. Also note that the values $\bar{x}$ and Sx are in this problem synonymous with $\bar{d}$ and Sd.

The test statistic can be calculated on the home screen by $t = (\bar{d} - \mu_0)/(Sd/\sqrt{n}) = (.2 - 0)/(2.387/\sqrt{(5)}) = 0.187$. This, of course, would require your knowing the values of $\bar{d}$ and Sd. These could have been found using 1-Var Stats from the STAT CALC menu for the differences in list L3.

Confidence Interval for μ_d

EXAMPLE Confidence Interval for Estimating Mean Weight Change: Use the same sample matched pairs to construct a 95% confidence interval estimate of μ_d which is the mean of the differences between April and September weights.

Press STAT, arrow to TESTS and choose Option 8 which is the TInterval. My TI-84 input screen is at right. The input for other models is analogous.

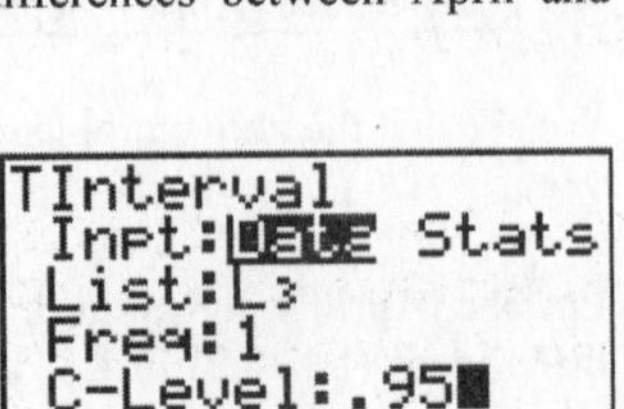

Press ENTER after highlighting Calculate on an 83/84 (OK on an Nspire). The 95% confidence interval is given as (–2.76, 3.16). Rounding, we can say we are 95% confidence that the mean difference in weight is between –2.8 and 3.2 kilograms. This implies that the mean difference could be 0 (because 0 is contained in the interval). Again, our sample has not given sufficient evidence that there is a difference in the means. We can find the margin of error as 3.2 – 0.2 = 3.0. The Home screen calculation of the margin of error of the above confidence interval would involve the formula $E = t_{\alpha/2}\frac{s_d}{\sqrt{n}} = 2.776*2.387/\sqrt{5} = 2.963$. The value 2.776 is from the t-table with (5-1) = 4 degrees of freedom.

COMPARING VARIANCES IN TWO SAMPLES

The following procedure is used to test if there is a difference in the variances of two populations based on two independent samples drawn from the populations. This test is sometimes used to decide whether or not to pool the standard deviations when comparing two population means. If this test determines the variances (and thus the standard deviations) are different then one would not want to pool the standard deviations. As always when working with variances, one must take extra care to ensure you are working with data from normal populations – this is a *very* critical assumption for this test.

EXAMPLE Comparing Variation in Weights of Quarters: Data Set 20 in Appendix B includes the weights (in grams) of samples of quarters made before and after 1964. In designing vending machines, we must consider the standard deviation of quarter weights so that coins are properly accepted. The sample statistics are summarized in the accompanying table. Use a 0.05 significance level to test the claim that the weights of pre-1964 and post-1964 quarters have the same standard deviation.

	Pre-1964	Post-1964
n	40	40
s	0.08700 g	0.06194 g

The above leads us to test the following hypotheses: H_0: $\sigma_1 = \sigma$ and H_1: $\sigma_1 \neq \sigma_2$. Your text shows histograms and normal quantile plots of the data; these indicate that these weights do come from normal populations, so the requirements for the test are met.

From the STAT TESTS menu, choose the option 2-SampFTest. The option number varies with the model of the calculator, but in all cases it is near the end of this menu. It is easiest to use the up arrow to locate it. Duplicate my input screen. (This time we are using the summary statistics instead of the raw data.)

NOTE: It is customary when using this test to use the larger standard deviation as the first sample.

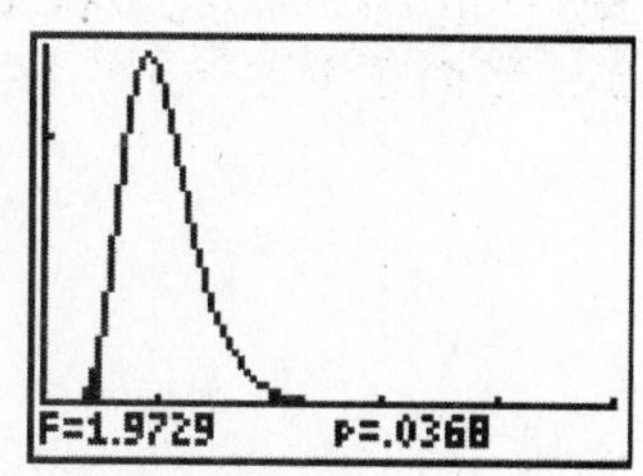

If you choose to highlight Draw and press ENTER, you will see the screen at right. The test statistic is F = 1.9729 and the p-value is 0.0368. One thing you can see from this graph is a major difference between F distributions and the normal and t distributions – F distributions are not symmetric. We conclude there is sufficient evidence to decide there is a difference in the standard deviations of the two types of quarters; it appears those made before 1964 are more variable in weight.

WHAT CAN GO WRONG?

Not Pairing Paired Data
This is a critical mistake. One needs to think carefully if there is some natural pairing of data that might (possibly) come from independent samples. Clearly, if the samples sizes are not the same, the data cannot have been paired. If one fails to pair paired data, wrong conclusions will usually be made, due to overwhelming variability between the subjects.

Bad Conclusions
The biggest thing to guard against is bad conclusions. Think about the data and what they show. Do not let conclusions contradict a decision to reject (this means we believe the alternate is true) or not reject (this means we have failed to show the assumed value is wrong) a null hypothesis. Remember, confidence interval gives a range of believable values for a parameter; because a claimed value is in the interval does not mean it is true – just reasonable. Also, one must be careful in keeping track of which sample was used as "group1" and which was "group2" in computing the test (and construction the alternate hypothesis).

Other than that, there is not much that hasn't already been discussed – trying to subtract lists of differing length will give a dimension mismatch error. Having more plots "turned on" than are needed can also cause errors.

10 Correlation and Regression

In this chapter we will see how TI calculators can help with simple linear regression and correlation. There are built-in functions on the STAT, CALC menu to aid in this work, but the most useful all-around is the LinRegTTest on the STAT TESTS menu. We will also study multiple regression. TI-83 and -84 calculators do not have a built-in function for this process, so on those calculators we will use a program called **A2MULREG,** included on the CD-ROM with your text and on the text's website. TI-89 and Nspire calculators have built-in multiple regression capabilities.

All TI-calculators compute the correlation coefficient as a part of linear regression, so we will not discuss that separately.

SIMPLE LINEAR REGRESSION AND CORRELATION

EXAMPLE Pizza Prices and Subway Fares: As described in the Chapter Problem, the price of a slice of cheese pizza and subway fares in New York City have seemed to cost the same since at least 1960. They also increase about the same time and by about the same amounts. The data are reproduced below.

	1960	1973	1986	1995	2002	2003
Cost of Pizza	0.15	0.35	1.00	1.25	1.75	2.00
Subway Fare	0.15	0.35	1.00	1.35	1.50	2.00
CPI	30.2	48.3	112.3	162.2	191.9	197.8

A Scatter Plot

Enter the data into L1 (Pizza) and L2 (Fare).

Set up Plot1 for a scatter plot as at right.

Press [ZOOM] [9] (on an 89, press [F5] after the plot has been defined) for the plot.

The scatter plot reveals a pattern that is relatively linear. There seems to be a strong relationship between the price of cheese pizza and subway fares.

Scatterplots on the TI-Nspire

The data have been entered in two lists that have been named subway and pizza. To graph the data, press [home] [5] for a new graph window. Move the cursor to the bottom of the window and press [click]. Select the variable pizza and you will have a dotplot of the pizza prices. Now, move the cursor to the left side of the window, press [click] for the list of named variables and select subway. The pizza price dots move upward to create the scatterplot.

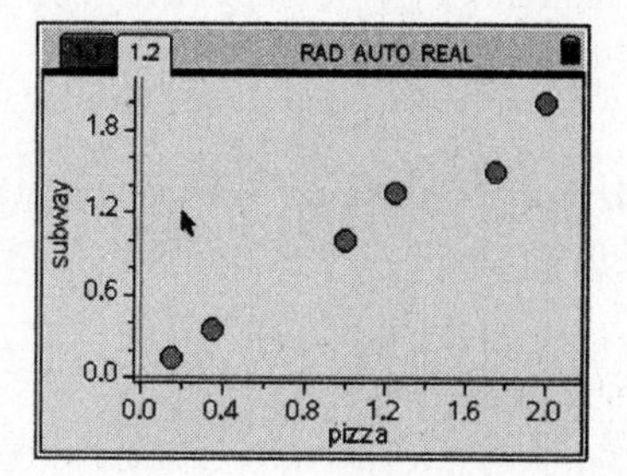

Linear Correlation Coefficient, r

EXAMPLE Pizza and Subways: Using the data, find the value of the linear correlation coefficient r.

Press STAT, arrow to TESTS and select LinRegTTest from the menu (the option number varies with calculator model, but is near the bottom of the list). Set up the screen as at right. On an 83 or 84, paste in Y1 from VARS ▶ 1 1 to store the regression equation for further use. (On an 89, use the right and down arrows to select a regression equation function.)

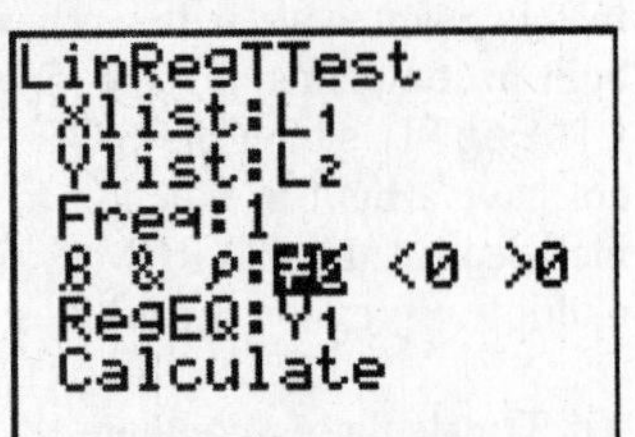

Highlight Calculate in the last line and press ENTER for the first of two screens of output. We are not interested in the t-statistic and its p-value at this point, so press the down arrow to scroll to the bottom.

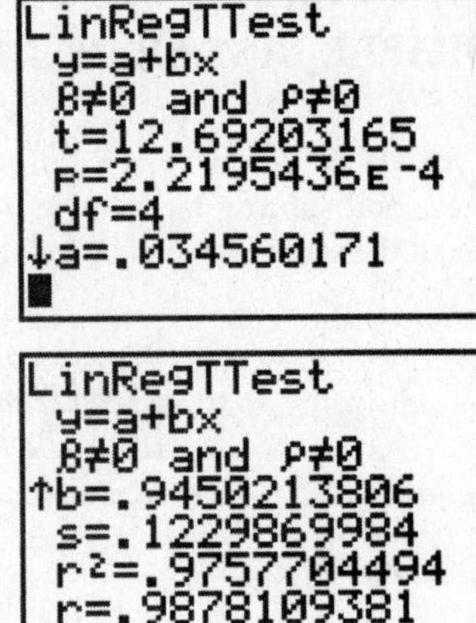

Note that the last line of output gives r = .9878109381. This indicates the linear relationship between these two variables is positive (as we saw in the graph) and very strong (because r is close to 1).

Note: There are other ways of calculating r on TI calculators, but this method is more useful all around as it provides many other results which will be of interest soon.

To perform the regression on a TI-Nspire, you can refer to the variables either by their column letter or by the variable name.

Explained Variation r^2 (Coefficient of Determination)

EXAMPLE Pizza and Subways: Referring to the pizza data, what proportion of the variation in subway fares can be explained by the price of cheese pizza?

In the screen above, we see that r^2 = .9577…, so about 95.8% of the variation in subway fares can be explained by the price of cheese pizza slices.

Formal Hypothesis Test of the Significance of r

EXAMPLE Pizza and Subways: Using the data above, test the claim that there is significant linear correlation between pizza prices and subway fares . This test of H_0: $\rho = 0$ versus H_1: $\rho \neq 0$ is exactly the same as a test of H_0: $\beta_1 = 0$ versus H_1: $\beta_1 \neq 0$.

Using the first output screen at the top of this page, we see the test statistic t = 12.692 and the p-value = 2.2195E-4 = 0.000222 < .05, indicating significant positive linear correlation (and a non-zero slope).

Regression

EXAMPLE Pizza and Subways: For the data we have been working with, find the regression equation of the straight line that relates pizza price and subway fares.

The equation has the general form $\hat{y} = b_0 + b_1x$. Your calculator uses the form $\hat{y} = a + bx$, so $a = b_0$ and $b = b_1$. You can look at the output above and see that a = 0.035 and b = 0.945 (rounded to three places as suggested in the text). Thus, your equation is $\hat{y} = 0.035 + 0.945x$. Be careful – many instructors will insist (like the author of this manual) that you report your regression equation *in terms of the actual variables with units* and *not* x and y. In this case, I would report "Subway Fare ($) = 0.035 + 0.945*Cheese Pizza Price ($)"

When we performed the calculation, we told the calculator to store the regression equation as Y1. You can check this by pressing [Y=]. ([♦][F1] = [Y=] on a TI-89.) It would be nice to see how the line passes through the data points.

We already defined the scatter plot of the data. If all other plots are turned off, you can press [ZOOM] [9] to see the regression line plotted through the scatter plot of points. (On a TI-89, go back to the plot set-up screen ([F2] from the Statistics/List Editor, and press [F5] to display the plot).

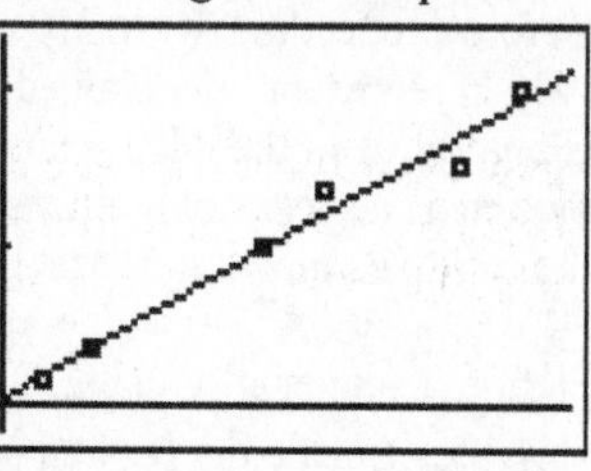

If you are using an Nspire, return to the graph window and press (menu), (4) for Analyze, (6) for Regression, and (2) for Show Linear a+bx. The regression line and its equation are added to the graph.

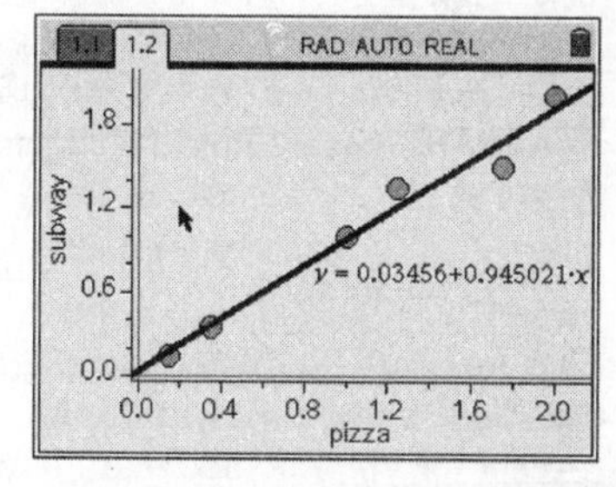

Predictions

EXAMPLE Pizza and Subways: Using the sample data, we found that there is a significant linear correlation between the price of pizza and the price of a subway fare in New York City. You also found the regression equation. Now suppose the price of a slice of cheese pizza has gone up to $2.25. Use the regression equation to predict the next subway fare.

Because $2.25 is not strictly within the range of our data (this is extrapolation), we'll need to resize the graph window and exercise caution with our prediction. Use [WINDOW] to increase Xmax to 2.3; then press [GRAPH] to display the resized plot. With the regression line and scatter plot displayed as above, press [TRACE] ([F3] on an 89) followed by [▼] to hop from a data point to the regression line. You can use the left and right arrows to try to locate x = 2.25 on the line, but it will be difficult.

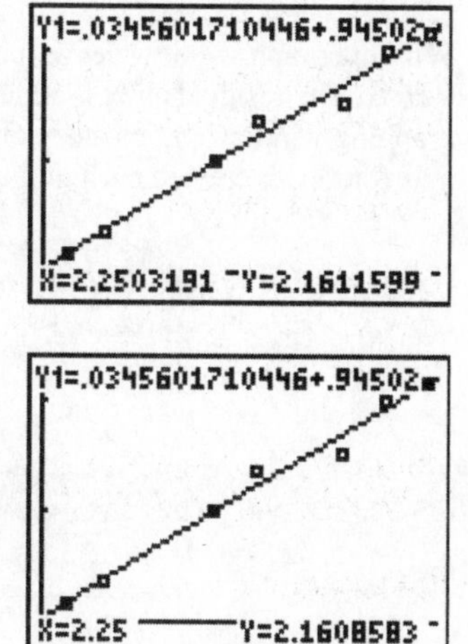

Type in 2.25. A large X=2.25 will show in the bottom line. Press [ENTER]. We see the predicted y value is $2.16 for a subway fare when cheese pizza slices are $2.25. If you are trying to do this on an Nspire, press (menu), then (4) for Analyze then (A) for Graph Trace. Nspires will not allow you to type in the *x* value explicitly here.

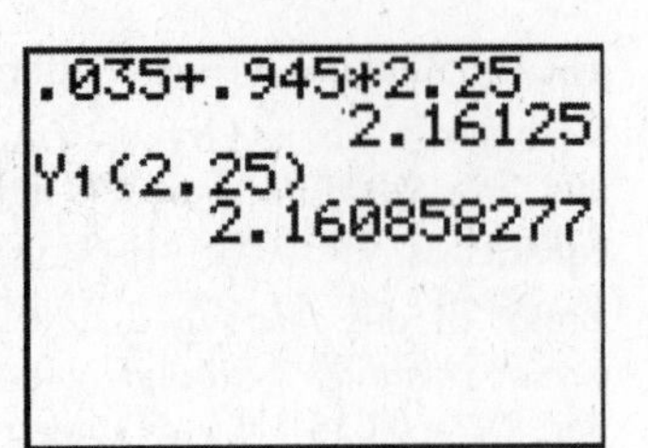

You can also calculate the predicted $\hat{y}$ on the Home screen by typing in the regression equation as in the top line of my screen or by pasting Y1 ([VARS] [▶] [1] [1]) to the Home screen and then typing (2.25) and pressing [ENTER], as in the last line of the screen. The function is saved as `f1` on the Nspire; otherwise using it is similar. Notice a small difference due to rounding. (On a TI-89, locate the `y1` variable from the [2nd][–] (`VAR-LINK`) screen.

Residuals Plots

Before proceeding with any more inference on a regression, it is good practice to examine a plot of the residuals (which represent unexplained variation around the line). We do this because these plots help assess whether the line is a good fit to the data. Having subtracted out the linear trend will magnify any remaining patterns. Ideally, when plotted as a scatter plot against the original *x* variable, one will see "pure, random scatter." Any noticeable pattern in the residuals plot means that the linear regression is *not* a good model for that set of data.

Define a scatter plot using the original *x* (predictor) variable, and `RESID` as the *y* variable. Locate the list name `RESID` on the [2nd][STAT] (`LIST`) menu of list names. TI calculators compute the residuals automatically when performing a regression. If you are using a TI-89, you will actually see these as a new list in the list editor after calculating the regression. Find the list name on the [VAR-LINK] screen in the `STATVARS` folder. Press [2] ('r' to move to the correct portion of the list).

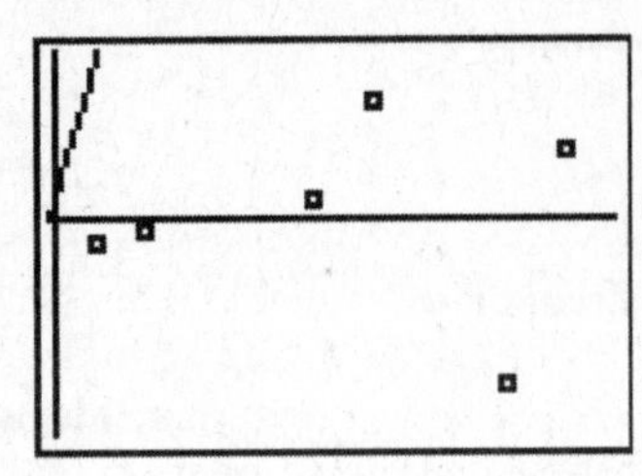

Press [ZOOM][9] ([F5] on an 89) to display the graph. Here we see pretty much random scatter, so the linear regression is adequate. One cautionary note: the calculator always tries to display everything it can – any lines seen in this plot (except for y = 0 in the center) like the diagonal at the left (the regression line) are not part of the plot!

NSpire Residuals

With the graph window active, press (menu), (4) for `Analyze`, (7) for `Residuals`, then (2) for `Show Residual Plot`. The residuals plot is added in a split window below the original data plot. You can also define a new graphing window and create another scatter plot with pizza on the x axis and the variable `stat.resid` on the y axis (these are created each time you do a regression).

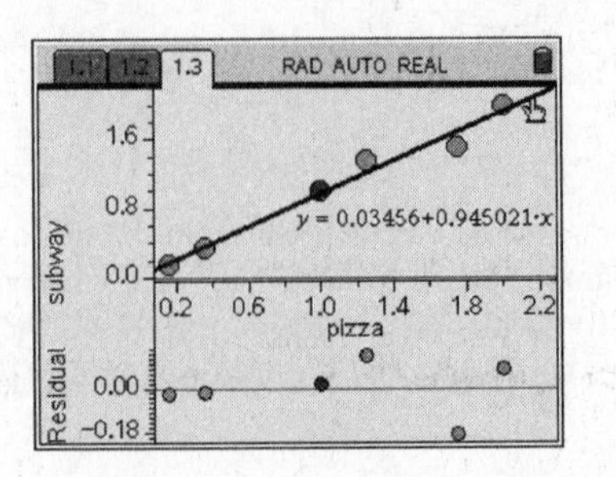

Prediction Intervals

We met confidence intervals for the mean of a distribution earlier. So far we have used our regression equation to find a point estimate of the response for a given x – but there is still variation around the linear model (represented by the residuals). We'd like a prediction *interval* for a particular value of the x variable – one that gives an idea of how precise our estimate is.

EXAMPLE Pizza and Subways: For the Pizza data, we just found that the best prediction for $x = 2.25$ (a \$2.25 slice of cheese pizza) is $\hat{y} = \$2.16$ for a subway fare. Find a 95% prediction interval for the price of a subway fare when the price of a pizza slice is \$2.25.

TI-83/84 Procedure

The only way to do this is to calculate the interval on the Home screen using the formula. From Table A-3 of the text (or using `invT` on an 84), we find the value $t_{\alpha/2} = 2.776$ using 6 - 2 = 4 degrees of freedom. The other quantities needed in the formula will be stored in the calculator after you have performed the `LinRegTTest`. You can paste them in from the VARS menu as described below.

First we store 2.25 as X, 2.776 as T and `Y1(X)` as Y. Here `Y1(X)` is the value we get when we input $x = 180$ into the regression equation stored in Y1. This can be done as separate commands as shown, or as one command with the individual parts separated by a colon ([2nd][.]). We find the predicted value on the line is \$2.16 for a cheese pizza slice price of \$2.25.

```
2.25→X
                2.25
2.776→T
               2.776
Y1(X)→Y
         2.160858277
```

Now we calculate the margin of error E using the formula from the text,

$$E = t^*_{\alpha/2,n-2}\sqrt{1+\frac{1}{n}+\frac{(x-\bar{x})^2}{n\sum x^2-\left(\sum x\right)^2}}.$$

Values used in the formula are stored in the Statistics submenu of the VARS menu. You can find and paste in:

- **s** at [VARS] [5] [▶] [▶] [▶] [0] (in the `Test` menu)
- **n** at [VARS] [5] [1] (in the XY menu)
- **x̄** at [VARS] [5] [2] (in the XY menu)
- **Σx²** at [VARS] [5] [▶] [2] (in the Σ menu)
- and **Σx** at [VARS] [5] [▶] [1] (in the Σ menu)

Press [ENTER] to reveal that E = 0.4416.

Note: As you work, look carefully at the quantities in the statistics submenus, so you can familiarize yourself with what is available to you and where it is located.

Now we start with our prediction value Y from the regression equation itself and add and subtract the margin of error to form the 95% confidence interval $\$1.72 < y < \$2.60$. We are 95% confident that a subway fare should be between \$1.72 and \$2.60 when a cheese pizza slice is \$2.25.

Note: The prediction interval for different x-values and different confidence levels can be easily calculated using the last-entry feature ([2nd] [ENTER]). Simply recall the top input lines beginning with where we stored our x value and change the inputs as needed and press [ENTER]. Then recall the equation to recalculate E.

TI-89/Nspire Procedure

Both of these calculators have a built-in function to compute confidence intervals for both the slope and mean response at a particular x value, as well as prediction intervals for a particular response at a particular x value. From the `(Confidence) Ints` menu, select option `7:LinRegTInt`. Input the names of the X List and Y list you are using, and indicate that each entry is a single observation. Select a regression equation to store the results. With the right arrow, you can select to compute an interval for either the slope or the Response. If for a response, you will then be allowed to enter the particular x value. Use the down arrow to enter the particular confidence level for your problem.

After pressing [ENTER], you will see the first portion of the results. Here, we have $\hat{y}$, the value from the regression equation, and the confidence interval for the *mean* price of a subway ride for *all* time points in which the price of a slice of cheese pizza is $2.25.

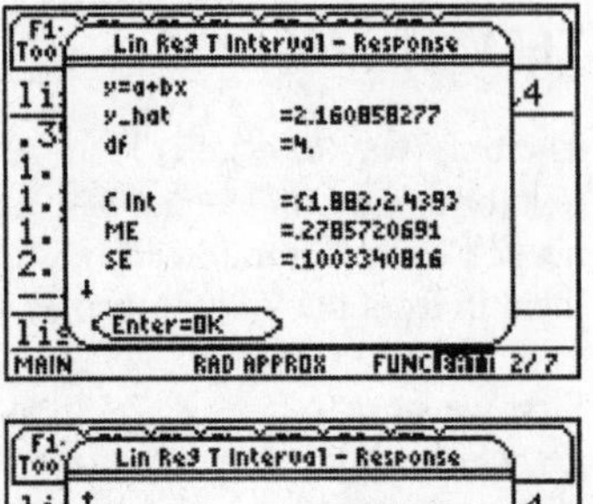

Using the down arrow, we find the prediction interval for a *particular* time when the price of a cheese pizza slice is $2.25. We are 95% confident that the price of a subway fare then will be between $1.72 and $2.60. Completing the output screen are the coefficients of the regression, r^2, r, and the x value used.

MULTIPLE REGRESSION

The TI-83 and -84 do not have built-in multiple regression capabilities (regression using more than one predictor variable). TI-89 and Nspire calculators do have built-in multiple regression. For the 83/84 calculators, we will use a program written for this purpose called **A2MULREG**. This program comes on the CD-ROM which accompanies your text. See the Appendix for details of transferring this program to your calculator. Program **A2MULREG** requires that the data set be stored in matrix [D] on your calculator. We will walk through this process below. One important restriction is that the dependent variable (Y) must be stored in the first column of [D]. This is not the case when using most statistical software such as Minitab (or the TI-89 and Nspire).

EXAMPLE Family Heights: Table 10-6 (reproduced below) gives the heights of mothers, fathers, and their daughters (in inches) for a random sample. Find the multiple regression equation to predict a daughter's height from her parents.

Mother	Father	Daughter		Mother	Father	Daughter
63	64	58.6		63	69	62.2
67	65	64.7		67	70	67.2
64	67	65.3		62	69	63.4
60	72	61.0		69	62	68.4
65	72	65.4		63	66	62.2
67	72	67.4		64	76	64.7
59	67	60.9		63	69	59.6
60	71	63.1		64	68	61.0
58	66	60.0		60	66	64.0
72	75	71.1		65	68	65.4

TI-83/84 Procedure:

The data can be imported from another TI-83 Plus or from a computer. The following method is for entering the data from a keyboard. Press 2nd x^{-1} ▶ ▶ to call the **MATRIX** menu and to choose the **EDIT** submenu. (If you are using a regular TI-83 calculator, it has a MATRX button to start the **MATRIX** menu.) Go down the screen to highlight [D] and press ENTER. You must first tell the calculator the size of the matrix you will be using. In the example below, we have 20 observations, and three variables, so the matrix is 20 x 3 (20 rows and 3 columns).

Now enter the data row by row; however, the dependent (Y) variable *must* be entered into the first column of the matrix. Start with the cursor in the first row, first column. Type each data set value across rows followed by ENTER. Press 2nd MODE to exit the editor.

With the data stored in [D], we now use the program **A2MULREG** to verify the Minitab output given in the text for the three most important components. When the program is first started, you will see this screen. An indication that the calculator is paused is given by the scrolling dots in the upper right corner. This screen is just a reminder of the data input requirements and informs you which other matrices will be used. If your data are not as they must be (or you want to save information in one of the other matrices, you can turn the calculator off and then back on to abort running the program.

Press ENTER to get the menu screen. Press ENTER to select a multiple regression.

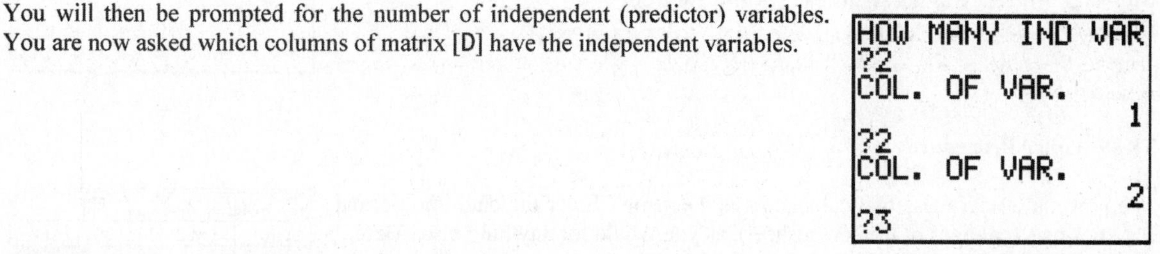

You will then be prompted for the number of independent (predictor) variables. You are now asked which columns of matrix [D] have the independent variables.

Here is the first screen of output. What we see first is the analysis of variance output for the general usefulness of the model. This F test tests the hypotheses H_0: all $\beta_i = 0$ versus H_1: not all $\beta_i = 0$. In this case, the F statistic is F = 17.67 with p-value p = 0.000, so we conclude that not all the coefficients of our predictor variables are 0 (at least one is significantly related to the daughter's height). We also see r^2 for the entire model, the r^2 adjusted for multiple predictor variables, and the standard deviation of the points around the regression "surface" (it's not a line in multiple dimensions).

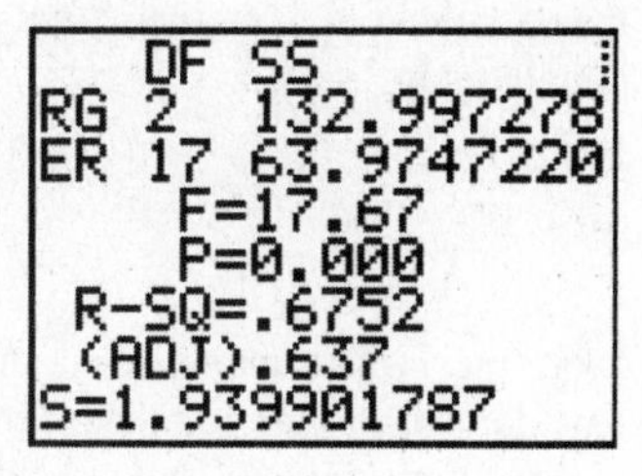

Press ENTER again to find the estimates for the coefficients, along with their individual *t* statistics and p-values.

The multiple regression equation is given as
Daughter's Height = 7.454 + .707*Mother + 0.164*Father.

We are also given the t-statistics for testing the individual coefficients and the p-values for the test of H_0: $\beta_i = 0$ versus H_1: $\beta_i \neq 0$. Notice that for the third column (the variable Father's height) the *t*-statistic is small and the p-value large, so this coefficient is not significantly different from 0.

Press ENTER again and you will be presented other choices to continue the analysis. The first option will compute confidence and prediction intervals for combinations of the predictor variables (the program will tell you the degrees of freedom for the model – you will need to supply the appropriate *t*-multiplier); the second will do residuals plots. The submenus and prompts for these options are self-explanatory.

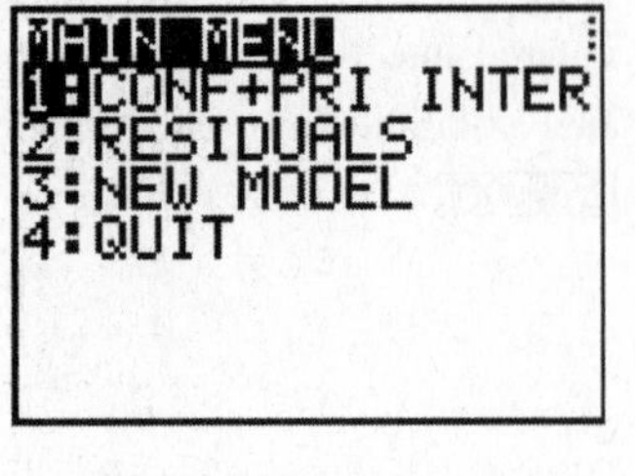

Correlation Matrix

EXAMPLE: What is the best regression equation if only a single independent variable is to be used?

When you ran Program **A2MULREG**, the first menu screen gave you the option of calculating the correlation matrix. Begin running the program again. This time highlight the option of calculating the correlation matrix.
Note: Pressing ENTER from the Home screen will restart the program if the last thing done was to quit it.

Press 2 and wait patiently as this calculation can take some time, depending on the number of variables and data points. The partial output is given at right. Looking at just the first column, we can see that the last variable, Father's Height, has a small linear correlation with the first variable, Daughter's Height. This explains why its coefficient was not significantly different from 0.

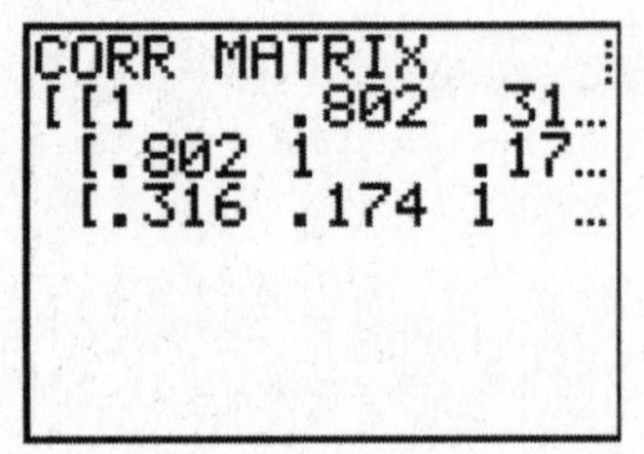

TI-89/Nspire Procedure

The procedure with these two calculators is the same. Enter the data into normal statistics lists (columns of the spreadsheet), as you would for anything else. Here, I have entered the mothers' heights in `list1`, the fathers in `list2`, and the daughters in `list3`.

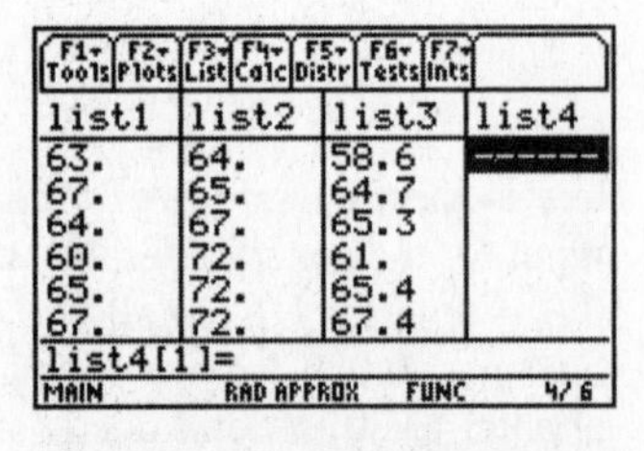

From the `Tests` menu, select option `B:MultRegTests`. You first must tell the calculator how many independent variables you will use. Use the right arrow key to open the submenu, and use the down arrow to find the appropriate choice. This will change the number of boxes available to specify independent variables. For our problem, there are 2 independent variables, which are in lists 1 and 2.

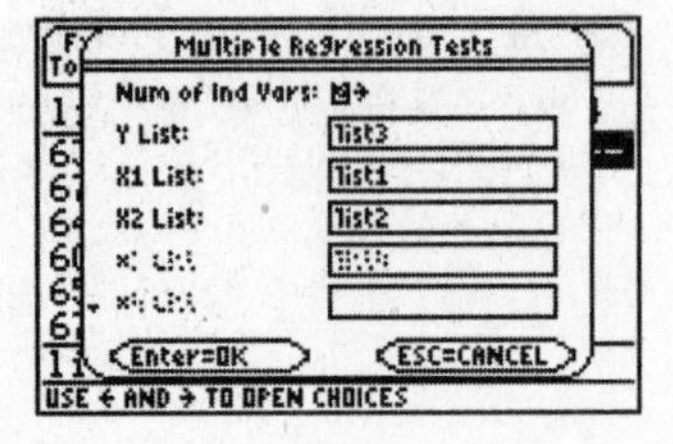

Press ENTER for the first portion of output. This gives the F statistic and p-value for the overall "utility" of the model. This tests the hypotheses H_0: $\beta_i = 0$ versus H_1: not all $\beta_i = 0$. It also gives the values of r^2 and r^2 adjusted for additional variables in the model. The standard deviation of the residuals is given, as is the Durbin-Watson statistic , DW, which measures the amount of correlation in the residuals and is useful for data which are time series (data that has been collected through time). If the residuals are uncorrelated, this statistics will be about 2 (as it is here); if there is strong positive correlation in the residuals, DW will be close to 0; if the correlation is strongly negative, DW will be close to 4. Since these data are not a time series, DW is meaningless for our example.

1.1 RAD AUTO REAL

	E	F
•		=MultReg
1	Title	Multiple R.
2	RegEqn	b0+b1*x1..
3	F	17.6707
4	PVal	0.000071
5	R²	0.675209

G1

Pressing the down arrow several times we find the components for regression and error which are used in computing the F statistic. The F statistic for regression is the MS(Reg)/MS(Error) where the Regression Mean square functions just like the treatment (factor) mean square in Analysis Of Variance (ANOVA).

1.1 RAD AUTO REAL

	E	F
•		=MultRegTests(c[],a[
9	dfReg	2.
10	SSReg	132.997
11	MSReg	66.4986
12	dfError	17.
13	SSError	63.9747

E13 ="SSError"

Finally we see some of the entries in new lists that have been created. On the TI-89, the complete lists are stored in the Statistics editor and will be seen when ENTER is pressed. On the Nspire, the complete list will be displayed by moving the cursor to the list entry spot in the spreadsheet. `Blist` contains the estimated intercept and coefficients; `SE list` is the list of standard errors for the coefficients which can be used to create confidence intervals for true slopes; `t list` gives values of the *t*-statistics for hypothesis tests about the slopes and intercept; `P list` gives the p-values for the tests of the hypotheses H_0: $\beta_i = 0$ against H_A: $\beta_i \neq 0$. If the assumed alternate is 1-tailed, divide these p-values by 2 to get the appropriate p-value for your test.

1.1 RAD AUTO REAL

	E	F
•		=MultRegTests(c[],a[
14	MSError	3.76322
15	bList	{7.4543048167295,0...
16	tList	
17	PList	{0.50250687339656...
18	SEList	{10.880368893642,0...

F15 ="{7.4543048167295,0.707205968082"

On a TI-89, after pressing ENTER we see all the new lists that have been added into the editor. `Yhatlist` is the list of predicted values for each observation in the dataset based on the model ($yhat_i = b_0 + b_1 x_{1i} + b_2 x_{2i}$ in this model); `resid` is the list of residuals $e_i = y_i - yhat_i$. `Sresid` is a list of standardized residuals obtained by dividing each one by *S*, since they have mean 0. If the normal model assumption for the residuals is valid, these will be N(0, 1).

yhatl...	resid	sresid	lever...
62.481	-3.881	-2.159	.14122
65.473	-.773	-.4382	.17308
63.679	1.6214	.86354	.06323
61.668	-.668	-.3796	.17713
65.204	.19602	.10634	.09703
66.618	.7816	.43116	.12676

leverage[1]=.141219121183...

MAIN RAD APPROX FUNC 10/15

Leverage is a measure of how influential the data point is. These values range from 0 to 1. The closer to one, the more influential (more of an outlier in its *x* values) the point is in determining the slope and intercept of the fitted equation. Values greater than $2p/n$, where n is the number of data points and p is the number of parameters in the model, are considered highly influential. Here, $n = 20$ and $p = 3$, so any value greater than 0.3 will designate an observation as highly influential. Paging down this list in our example, we find that the tenth data point (the one with a mother's height 72 inches and father's height 75 inches with daughter's height 71.1 inches) is influential.

After pressing the right arrow, we find more lists. Cook's Distance in the next column is another measure of the influence of a data point in terms of both its *x* and *y* values. Its value depends on both the size of the residual and the leverage. The ith case can be influential if it has a large residual and only moderate leverage, or has a large leverage value and a moderate residual, or both large residual and leverage. To assess the relative magnitude of these values, one can compare them against critical values of an F distribution with p and $n - p$ degrees of freedom or use menu selection `A: F Cdf` from the F5 (`Distr`) menu. The largest value in the list is .433, so that point is not highly influential. `Blist` is the

cookd	blist	selist	tlist
.25541	7.4543	10.88	.68512
.0134	.70721	.12886	5.4882
.01678	.16363	.12659	1.2926
.01034	------	------	------
.00041			
.009			

tlist[1]=.68511508107832

MAIN RAD APPROX FUNC 14/15

list of coefficients. We finally have the information we need to construct the fitted regression equation as: $\textit{Daughter's Height} = 7.454 + 0.707 * \textit{Mother's Height} + 0.164 * \textit{Father's Height}$. We interpret the coefficients in the following manner: A daughter's height increases by 0.707 inches for each inch of her mother's height, for fathers of the same height, while a daughter's height increases 0.164 inches for each inch of father's height, for mothers of the same height, on average.

The next column contains the standard errors of each coefficient. These can be used to create confidence intervals for the true values using critical values for the t distribution for $n - p$ degrees of freedom. Finally we see the t statistics and p-values for testing H_0: $\beta_i = 0$ against H_A: $\beta_i \neq 0$. These suggest the coefficient of father's height is not significantly different from 0; in other words, mother's height is a much more determining quantity for a daughter's height.

Confidence and Prediction Intervals

The TI-89 and Nspire can also automatically calculate confidence intervals for the mean response for a combination of the predictor variables, as well as prediction intervals for individual new observations. Use option 8:MultRegInt from the [F7] Ints menu. On the TI-89, specify the values for the variables in the X Values List box enclosed in curly braces ([2nd][(] and [2nd][)]) and separated by commas. On the Nspire, these values must be in a column of the spreadsheet; refer to the column letter or name.

Here, I have entered 66 inches for a mother's height and 74 inches for a father's height (the author and her husband's heights) in column D of the spreadsheet. The first page of output says the estimated average daughter's height for parents with these heights is 65.9 inches; the lower end of the confidence interval is 64.6 inches; paging down, the upper end of the confidence interval is 67.3 inches. The average height of *all* daughters for parents with these heights should be between 64.6 inches and 67.3 inches. The Prediction interval is wider; this gives a predicted height for *a particular* daughter of one set of parents with these heights. This daughter should be between 61.6 and 70.2 inches tall with 95% confidence (both of our daughters heights are in this range!).

Multiple Reg Intervals
X2 List: b
X Values List: d
Save RegEqn to: f4
C Level: 0.95
1st Result Column: e[]
OK Cancel

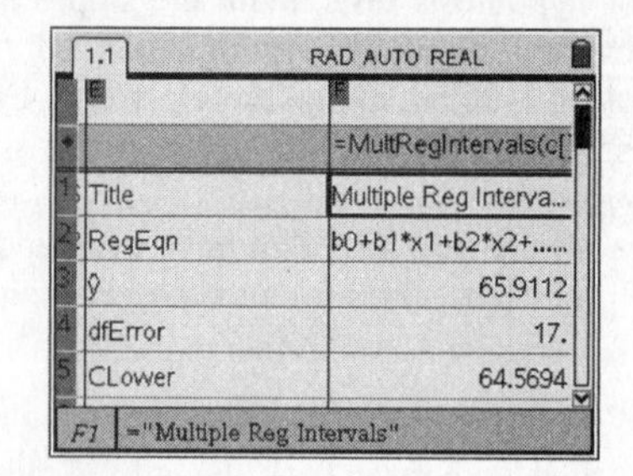

1.1 RAD AUTO REAL
=MultRegIntervals(c[
CUpper 67.253
ME 1.34182
SE 0.635991
LowerPred 61.604
UpperPred 70.2184
F10 =70.218368258696

How do I get rid of those extra TI-89 lists?

From the Statistics list editor, press [F1] (Tools) and select option 3:Setup Editor. Performing this action (leave the input box blank) recovers any deleted lists and deletes any calculator-generated lists.

11 Goodness-of-Fit and Contingency Tables

In this chapter we will use the TI-83 Plus STAT Editor as a spreadsheet to calculate the χ^2 test statistic used in the goodness of fit test for multinomial experiments. If you are using one of the other models, this is a built-in function. We will also use a built-in test from the STAT TESTS menu to do contingency table analysis.

MULTINOMIAL EXPERIMENTS: GOODNESS OF FIT TESTS

EXAMPLE Last Digits of Weights: The table below is a recreation of part of Table 11-2 in your text. It contains a frequency analysis of the last digits of weights obtained from 40 randomly selected adult males and 40 adult females, obtained as part of the National Health Examination Survey. By examining the last digits, researchers can verify that the subjects were actually weighed (instead of merely asking their weights, as these are usually rounded to multiples of 5). Test the claim that sample weights do not have the same frequency. Use significance level α = 0.05. What can we conclude about the procedure used to obtain these weights?

Last digit	0	1	2	3	4	5	6	7	8	9
Expected Frequency	.10	.10	.10	.10	.10	.10	.10	.10	.10	.10
Observed Frequency	7	14	6	10	8	4	5	6	12	8

Calculating Expected Frequencies (all models)

Put the observed frequencies into L1 and the expected proportions into L2.

Highlight L3 and type L2*sum(L1) as in the bottom of the screen (Find the sum(function at 2nd STAT ▸ ▸ 5, which is option 5 on the LIST MATH menu.)

Press ENTER for the expected frequencies (E) in L3. (If these values are given or are easy to calculate as in this example, you can simply enter them into L3.)

Conducting the χ^2 Test – TI-83 Procedure

Highlight L4 and type (L1-L3)²/L3. Press ENTER for the contribution made by each of the digits to the overall chi-square statistic.

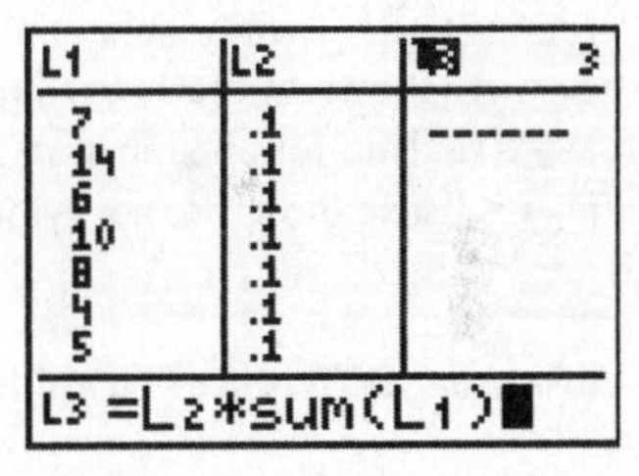

We can see that the largest contribution is made by an ending digit of 1 (the contribution is 4.5). The next largest contribution is from 5. If people were merely reporting their weights, we'd expect to see lots of 5's in the data; here, we actually had 14 individuals with last digit 1 when we expect 8; 5's were underrepresented: there were only 4 of those in our data.

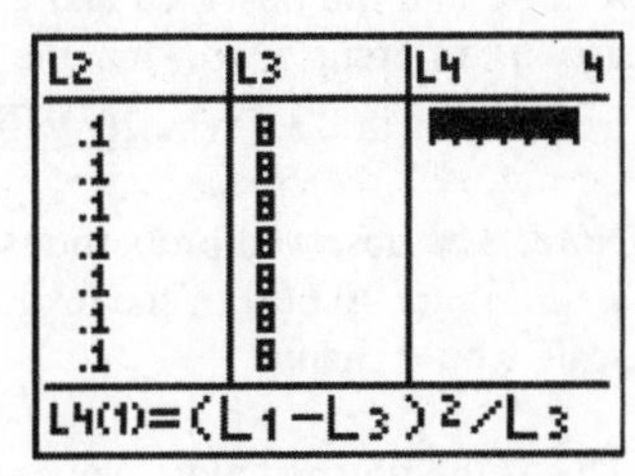

Press [2nd] [MODE] to Quit and return to the Home Screen. Sum the elements of L4, using the Sum(command from the LIST MATH menu (as above). We see the chi-square statistic is 11.25.

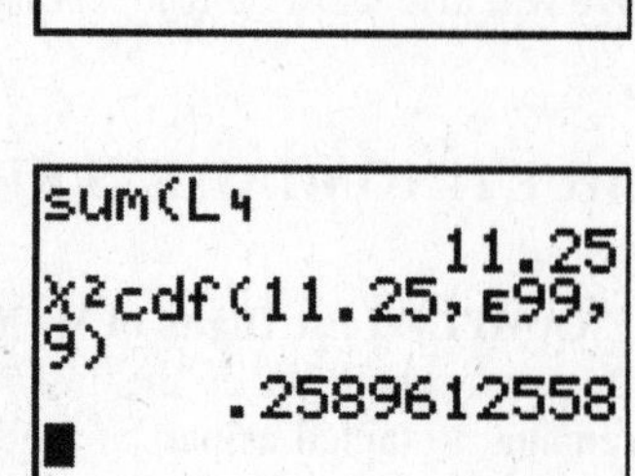

Computing the p-value of the χ^2 Statistic

Press [2nd] [VARS] [7] to paste χ^2 cdf(on the home screen. Then type 11.25,E99,9 and press [ENTER] for a p-value as shown. (You have asked for the probability of a test statistic greater than 11.25 in a chi-square distribution with df = 9.) Since the p-value is 0.259 and is much larger than the significance level $\alpha = 0.05$, we would not reject the null hypothesis that the weights were actually measured. Any observed differences in the last digits from what was expected were due to randomness.

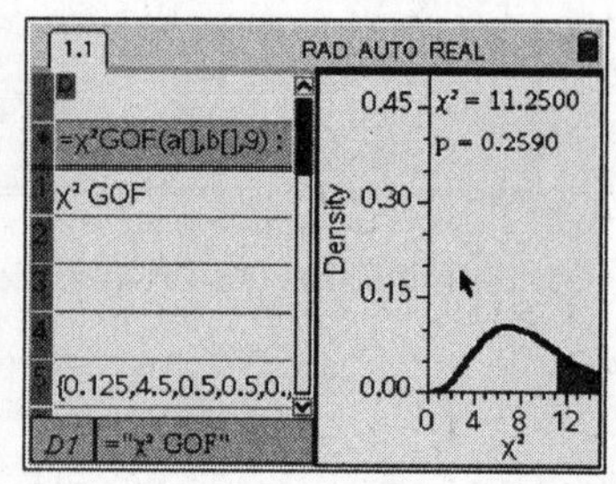

Conducting the χ^2 Test – TI-84/89/Nspire Procedure

All of these models have a built-in function to compute the test statistic and find the *p*-value of the test. Select χ2GOF from the Stat Tests menu. All models ask for the observed and expected (count) lists, as well as the degrees of freedom (categories – 1). You also have your choice to either simply calculate or draw the results.

Pressing [ENTER] will display the results. The TI-89 and Nspire will also create a new list (shown at the lower left in this Nspire output) with the components of the χ2 statistic for examination purposes. This Nspire output illustrates the fact that χ2 distributions are not symmetric.

Observing Differences Graphically

To graphically compare the observed frequencies with the expected frequencies, put the integers 0 to 9 in L5.

Set up Plot1 and Plot2 to be connected scatterplots using the digits in L5 as the Xlist and the observed and expected frequencies as the Ylist. As illustrated, use a different mark for the observed frequencies in L1 and the expected frequencies in L3. Press [ZOOM][9] [TRACE] to view the plot.

STAT PLOTS
1:Plot1...On
L5 L1
2:Plot2...On
L5 L3
3:Plot3...Off
L6 L4
4↓PlotsOff

Note: The observed proportions and expected proportions could have been plotted as in Figure 10.6(a) of the text, but the graph would look the same-only the y-axis scale would change.

This is the finished plot. Notice last digits 1 and 8 are over-reported, we would not expect this if the people had not actually been weighed.

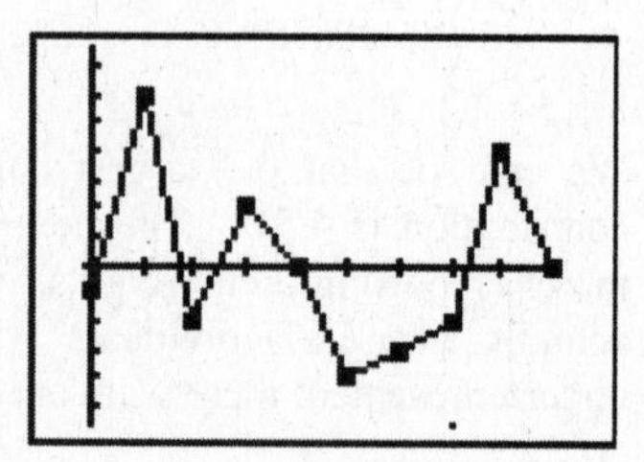

Our conclusion about the weights from this study: the individuals were actually weighed, not merely asked to give their weights.

EXAMPLE World Series Games: Below, we repeat Table 11-4 which lists the number of games placed in the baseball World Series, as of 2007. The table also lists the expected proportions for the number of games in the series, assuming both teams have about the same chance of winning. Use a significance level $\alpha = 0.05$ to test the claim that the actual number of games fit the distribution indicated by the probabilities.

GAMES PLAYED	4	5	6	7
Actual Contests	19	21	22	37
Expected Proportion	2/16	4/16	5/16	5/16

Here, I have entered the observed frequencies into L1, the expected frequency (you can enter these as the fractions given) in L2, and computed the expected frequency (count) in L3, just as was done above (multiply L2 by the sum of L1). We note that there were more four- and seven-game series than expected, and fewer six-game series.

On my TI-84 calculator, I have completed the input screen, similarly to that done above on an Nspire. This time, I have chosen the Draw output option. Be sure to have turned off all STAT PLOTS if you use this option (an error will most likely happen).

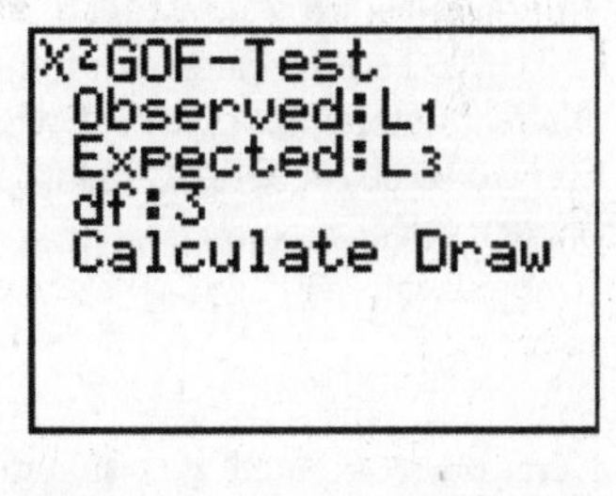

The distribution curve displayed is not symmetric. Further, (to 4 decimal places) the p-value of this test is 0.0485. The computed test statistic is 7.8848. At the 0.05 level, the actual World Series results do not fit the expected number of games, assuming the teams are equally matched. According to your text, so far, no reasonable explanations have been provided for the discrepancy.

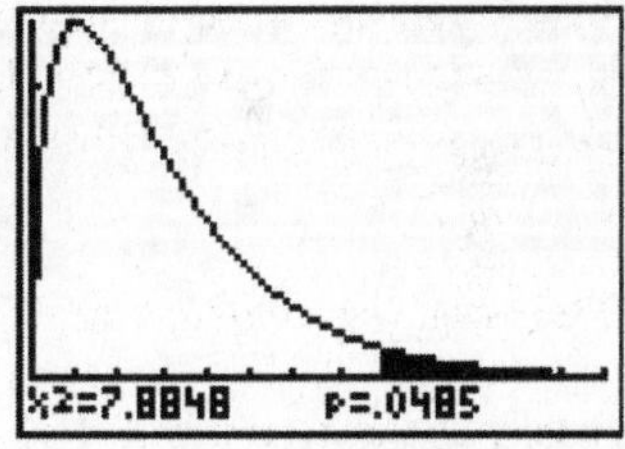

CONTINGENCY TABLES

Contingency tables are used to analyze count data for two types of relationships: homogeneity and independence. The test is the same for both – a $\chi 2$ test, but the setting and conclusions are very different. It is important to make the connection about what type of test you are performing. A test of independence usually involves the same individuals or items categorized by two variables. We want to know whether there is an association (or not) between the two variables. A test of homogeneity involves a single variable observed at different times, or in different populations. Here, we want to know if the distribution of the variable is the same in each.

EXAMPLE Echinacea and Colds: The data presented in Table 11-6 represent the results of an experiment about the effectiveness of Echinacea in treating colds. All the participants in the study were exposed to the rhinovirus after being treated with various levels of Echinacea. We are interested in the number who developed colds. Use a 0.05 significance level to test the claim that getting an infection (cold) is independent of the treatment group.

	TREATMENT		
	Placebo	Echinacea: 20% extract	Echinacea: 60% extract
Infected	88	48	42
Not infected	15	4	10

TI-83/84 Procedure

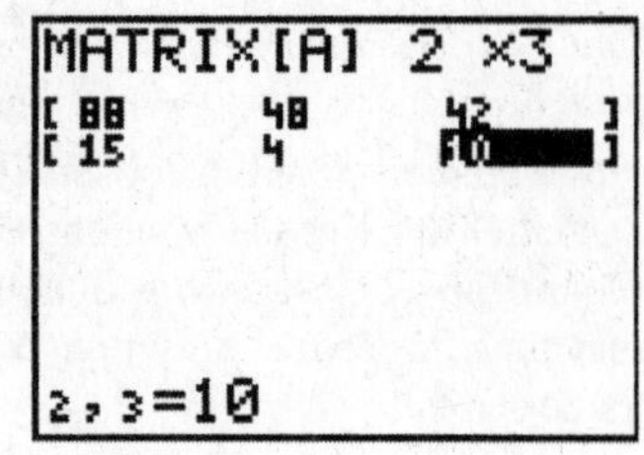

We must first enter the contingency table into a matrix. Press [2nd][x^{-1}] ([MATRX] on a regular TI-83) and arrow to EDIT. Select a matrix (usually [A]) and press [ENTER] to get the input screen. You first must tell the calculator the size of the matrix. Ours has two rows and three columns, so press [2][ENTER][3][ENTER] to set the dimensions. Now input the data (across rows) pressing [ENTER] after each. Press [2nd][MODE] to Quit the editor.

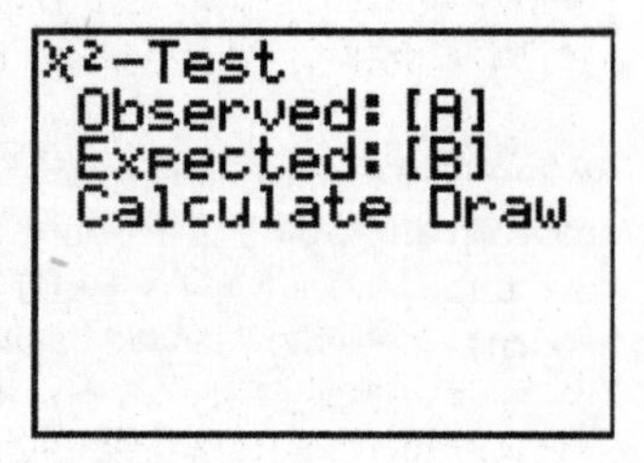

Press [STAT], arrow to TESTS, and select C:χ2-Test. The input screen is pretty obvious. Tell the calculator the name of the matrix with your observed counts, and where to store the matrix it will create with the expected counts. Choose either Calculate or Draw, as usual. If you want to change a matrix name, press [2nd][x^{-1}] to access the menu of matrix names.

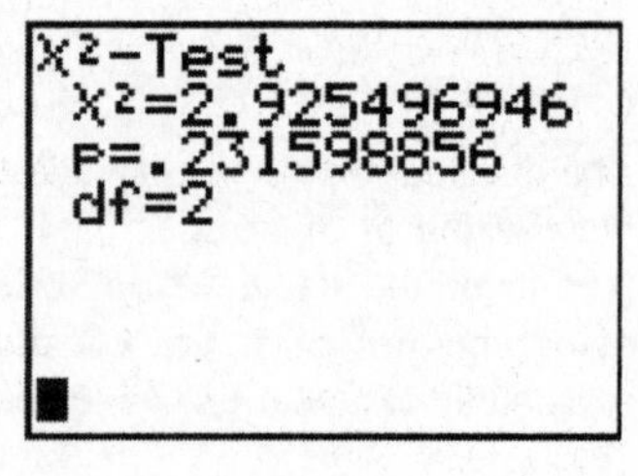

Here, our computed χ2 test statistic is 2.925, with p-value 0.2316. It appears that Echinacea is not effective in preventing colds.

TI-89 Procedure

On a TI-89, we must first select the Data/Matrix Editor application from the list of applications that display when the [APPS] key is pressed. The first thing you will see is this. It asks whether you wish to open a current problem, an existing problem, or start a new one. I have selected to begin a new problem.

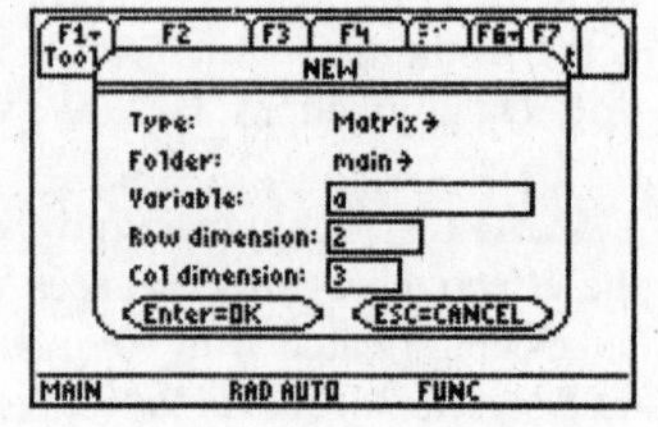

This next screen is used to define the type of entity you will be using. We want a matrix, so use the right arrow to expand the first type option box and use the down arrow to select (highlight) Matrix. You are asked which folder to store the matrix in; I recommend using main. Next, enter a name for the matrix, then the row and column dimensions. Press [ENTER] to proceed to the input screen.

Simply type in the entries (across the rows) and press [ENTER] after each. To exit the editor, press [2nd][ESC].

Return to the `Stats/List Editor`, and select option `8:Chi2 2-way` from the [F6] `Tests` menu. You are asked the name of the input matrix that you just created (access the name from [VAR-LINK] just as with list names), and have the option of changing the default names for the matrices of expected counts and components of the $\chi 2$ test statistic. Last, as usual, you have the option of `Calculate` or `Draw` for the results.

The results screen is at right. We see the value of the test statistic and its p-value along with degrees of freedom. We also see the first entry in both the Expected Value matrix and Components matrix.

To see the actual expected values or components matrices go to the home screen and press [2nd][−] ([VAR-LINK]). Both of these are stored in the `Statvars` folder. To find them more easily, highlight `Main` and press ◀ to collapse the entries in that folder. Move the cursor to highlight `Statvars`. If needed, press ▶ to expand that list. Locate the entry `compmat` and press [ENTER] to transfer the name to the home screen entry area, then press [ENTER] again to display the matrix. The largest contributor to the $\chi 2$ test statistic was for the cell for 20% Echinacea extract and colds.

Here, I have in a similar fashion displayed the entries in `expmat`, the matrix of expected counts. Comparing these to the matrix of original data, it seems that those who took the 20% extract actually had fewer colds (4) than expected (7); note that all the other cells have expected counts similar to the observed.

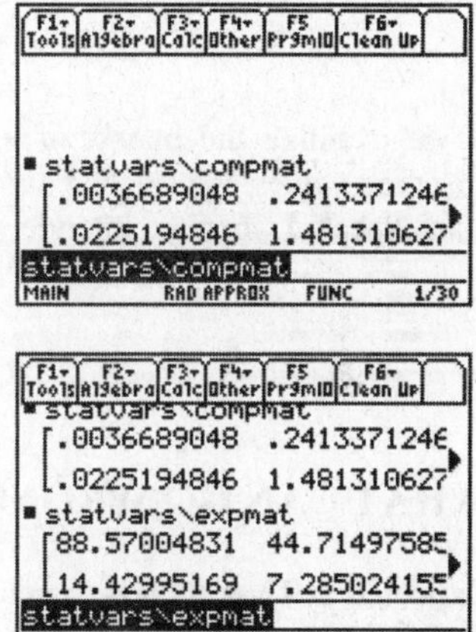

TI-Nspire Procedure

To create a matrix on the Nspire and input data, insert a blank Calculations page in a document, then press [ctrl][×] to insert blank square brackets. Type in the elements in the first row of the matrix, separating them with commas. Use a semicolon (find this in page 3 of the [ctrl][catalog]) and store the matrix with a name. The screen at right creates our observed count matrix and names it `a`. When you press [enter], it will display looking like a matrix. Select option $\chi 2$ `2-Way Test` from the `Statistics Tests` menu; use matrix `a` as input.

Tests of Homogeneity

In a test of homogeneity, we test the claim that different populations have the same proportions of some characteristic of interest. The procedure is the same as for tests of independence; one needs to be careful to identify the difference between the two.

EXAMPLE Influence of Gender: Does the gender of an interviewer affect the response given? A sample of 1200 men was asked if they agreed with the statement "Abortion is a private matter that should be left to the woman to decide without government intervention." The data in Table 11-8 are reproduced below.

	Gender of Interviewer	
	Male	Female
Agree	560	308
Disagree	240	92

I have entered the observed data into matrix [A], as shown at right.

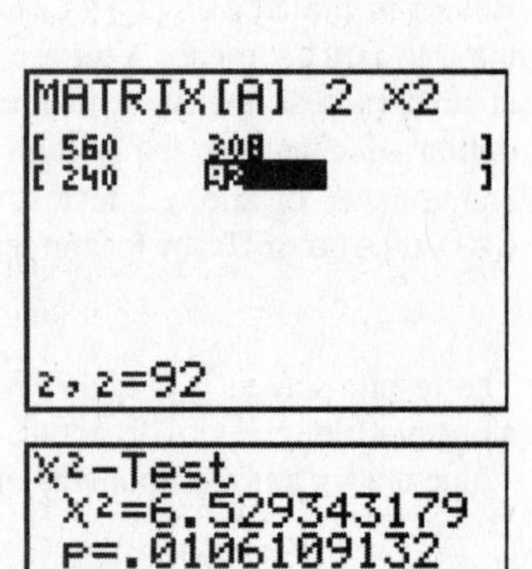

As detailed above, I have performed the $\chi 2$ test. The test statistic value is 6.529, with p-value 0.0106. We (at α = 0.05) will reject the null hypothesis that the proportion of those agreeing is the same for both genders of interviewers.

If we examine the matrix of expected values and compare it to the matrix of observed values, we see that our male subjects agreed with the statement less often than expected when the interviewer was a man.

WHAT CAN GO WRONG?

Expected cell counts less than 5.
Check the computed matrix of expected cell counts. If they are not all greater than 5 the analysis may be invalid. One way to handle these is to combine rows or columns in some logical way so that expected counts are at least 5.

Missing or misplaced parentheses.
When computing elements for the goodness-of-fit test on a TI-83, the parentheses are crucial.

Overusing the test.
These tests are so easy to do and data from surveys and such are commonly analyzed this way. The problem that arises here is that in this situation the temptation is to check many questions to see if relationships exist; but performing many tests on *dependent* data (the answers came from the same individuals) such as this is dangerous. In addition, remember that, just by random sampling, when dealing at α = 5% we'll expect to see something "significant" 5% of the time when it really isn't. This danger is magnified when using repeated tests - it's called the problem of multiple comparisons.

12 Analysis of Variance

In this chapter we will use the TI calculators' built-in function for doing one-way ANOVA problems. We will use a program called **A1ANOVA** (included on the CD-ROM) to extend our capabilities to do two-way ANOVA problems for two factor designs with equal numbers of observations in each cell on TI-83 and -84 calculators. The TI-89 and Nspire have built-in functions for this. The output for this program matches the Minitab ANOVA tables in the text. See the text for the proper interpretation for these tables.

ONE-WAY ANOVA

EXAMPLE Chest Deceleration in Car Crash Tests: Do car crash dummies tell us that larger cars are safer? In one test, cars are crashed into a fixed barrier at 35 mph with a test dummy in the driver's seat. Table 12-1 (below) lists chest deceleration measurements (in g, where g is a force of gravity). Larger values indicate greater forces of deceleration, which are likely to result in greater injuries to human drivers.

Small Cars (L1)	44	43	44	54	38	43	42	45	44	50
Medium Cars (L2)	41	49	43	41	47	42	37	43	44	34
Large Cars (L3)	32	37	38	45	37	33	38	45	43	42

Enter the data in lists L1 to L3 as indicated in the table.

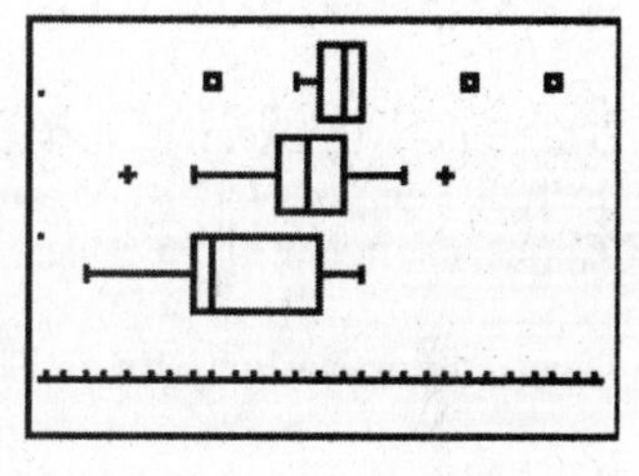

Plot the data. If there were more than three groups of data, the TI-83 and TI-84 calculators could not graph it all at once (TI-89 and Nspire can.) I have defined three modified boxplots to display the distributions of the decelertions for the various size cars. See Chapter 3 if you need to refresh your memory on these. Looking at the plots, we can see a decrease in the median chest deceleration with the increasing size of the car. Our plots indicate there are some potential outliers in both the small and medium size car data sets; however, none of the values are particularly unreasonably large or small (these data just seem to have a compressed range).

On a TI-83 or -84, press STAT ▸ ▸ ▴ ENTER to choose the last option on the STAT TESTS menu to paste ANOVA(on the Home screen. Type L1, L2, L3 as at right. On a TI-89 or Nspire, select option Anova from the Tests menu. The input and outputs are so obviously like what is shown, that no separate procedure to describe these is needed.

Press ENTER for the first portion of the results. We see the test statistic F = 4.094 and the p-value = 0.0280 just as seen in the text. The sum of squares for the factor (group) is shown along with its degrees of freedom and mean square. This small p-value confirms what our eyes already saw – the different size cars result in different chest decelerations. This does not, however, indicate which mean(s) is (are) different from the others – simply that one (or more) are different. Arrowing down, we see the same information given for the error sum of squares, and the (pooled) standard deviation.

Note: The program **A1ANOVA**, introduced in the next section, gives the means and standard deviations of the raw data stored in the lists. The program also accepts sample summary statistics (means, standard deviations, and sample sizes) as an input option in addition to the raw data option. The TI-89 and Nspire also accept summary statistics and have the capability to perform two-way ANOVA.

Which mean(s) is (are) different?

Having rejected the null hypothesis, we would like to know which mean (or means) is (are) different from the rest. For a rough idea of the difference one can do confidence intervals for each mean (using the pooled standard deviation found in the ANOVA) and look for intervals which do not overlap, but this method is flawed. Similarly, testing each pair of means (doing three tests here) has the same problem: the problem of multiple comparisons. If we constructed three individual confidence intervals, the probability (before data are collected) they all contain the true value is $0.95^3 = 0.8574$ using the fact that the samples are independent of each other.

The Bonferroni method is one way to make the distinction. In this, we must be careful to control for the number of possible tests. `1-Var Stats` tells us that the mean for L1 (small cars) is 44.7 and the mean for L2 (medium cars) is 42.1. Are these means different, or similar? We will compute a t-test for equality of these means, and use a method to adjust the p-value to account for the possibility of several tests. The t-statistic is $t = \dfrac{\bar{x}_1 - \bar{x}_2}{\sqrt{MSE\left(\dfrac{1}{n_1} + \dfrac{1}{n_2}\right)}}$. I have computed it at right, and found the one-sided p-value for a test of equality of means. Multiply by two for a two-sided test, and we have p = .2034. Those two groups have similar means. (Continuing, we find that large cars are different from small cars — merely confirming what our eyes already told us.)

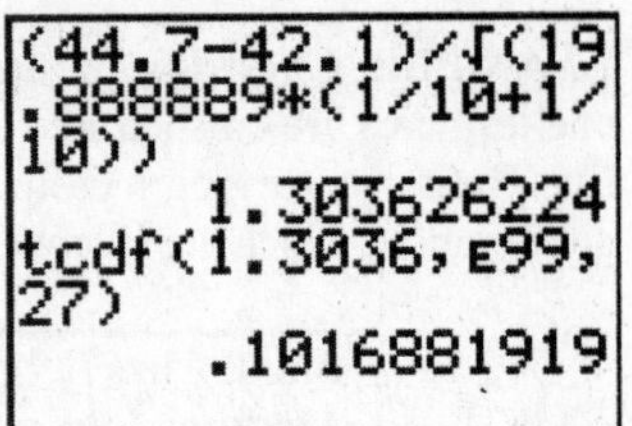

TWO-WAY ANOVA

The following example uses program **A1ANOVA** to perform two-way ANOVA. The program **A1ANOVA** is included on your CD-ROM, and is also available from the book's website. To download the program requires TI-Connect software and a cable.

Two-Factor Design with an Equal Number of Observations per Cell

EXAMPLE Foreign vs Domestic Car Saftey: We extend the example of the previous section to consider another question: is the difference in car safety similar in domestic and foreign-made cars?

	Small Cars	Medium Cars	Large Cars
Foreign	44 54 43	41 49 47	32 45 42
Domestic	43 44 42	43 37 34	37 38 33

The data in the table above will be stored in matrix [D] with 18 rows and 3 columns. The data values are all in the first column. The second column identifies the origin associated with each car (1=Foreign, 2=Domestic). The third column of [D] identifies the size of car (1 = Small, 2 = Medium, 3 = Large). Here are the step by step instructions.

Entering Data into Matrix [D]

Press [2nd] [x^{-1}] [▶] [▶] [4] to call the MATRX menu, the EDIT submenu and choose matrix [D] to edit. Type 18 [ENTER] 3 [ENTER] for 18 rows and 3 columns. Your matrix may not contain all zeroes, but that is fine as we will be typing over the values. Enter data row by row. Begin with 44 [ENTER] 1 [ENTER] 1 [ENTER]. Then continue typing in all the data values with their correct origin and size identifier. Press [2nd][ESC] to exit the matrix editor.

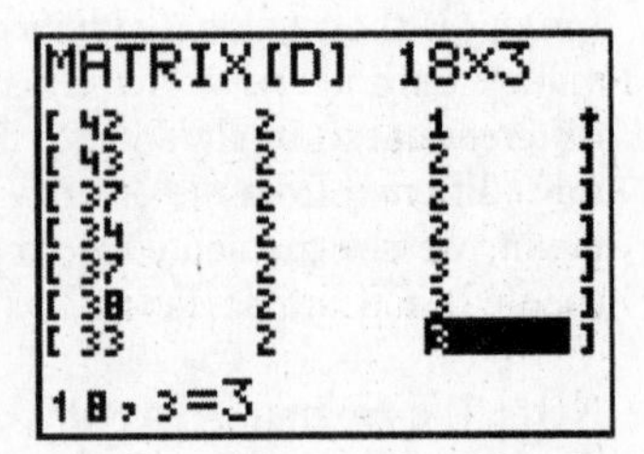

Running Program A1ANOVA

Press [PRGM] and highlight **A1ANOVA**. Press [ENTER] and prgm**A1ANOVA** is pasted to the Home screen. Press [ENTER] for the main menu screen. Press [3] to choose option 3, the 2WAY FACTORIAL.

You will then see this instructional screen. The purpose of this screen is primarily to remind you of the input data requirements in case you should forget. Press [ENTER] for another menu screen. This gives you two options: CONTINUE or QUIT. This allows you to quit if the data are not already in matrix [D] as required.

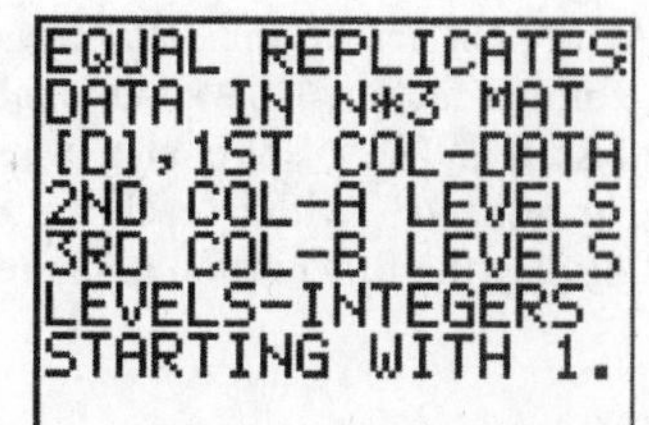

Press [1] to continue since our data is stored in [D]. The first part of the ANOVA table is shown on the screen. For each factor (A is the factor used in column 2 of the input matrix, so foreign or domestic for our example; B is for the car sizes), as well as interaction and error, the degrees of freedom and sum of squares is given. Mean squares for the ANOVA table are the sum of squares divided by the degrees of freedom. Next we have the F statistic for factor A (origin) and its p-value. With a p-value of 0.038, the origin does make a difference in chest deceleration (at least based on this small sample). The row of dots at the upper right indicates the calculator's output has been paused.

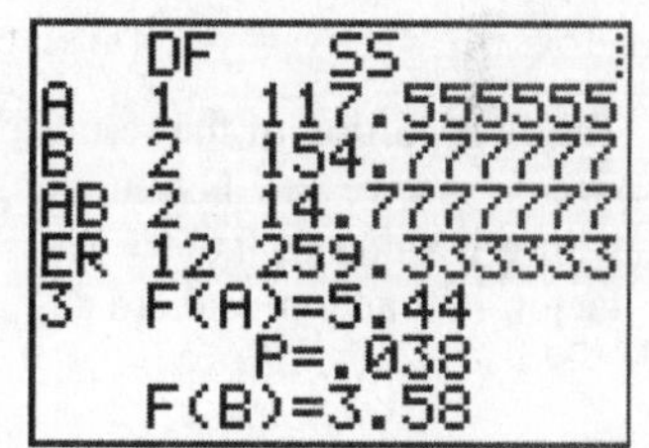

Press [ENTER] again for the rest of the output. The p-value for car size is 0.060. It seems car size is somewhat less important than country of manufacture. Last is the F statistic and its p-value for interaction. There is no interaction in our car crash data between origin and size. Lastly, the (pooled) standard deviation is given. Press [ENTER] to return to the first, main menu.

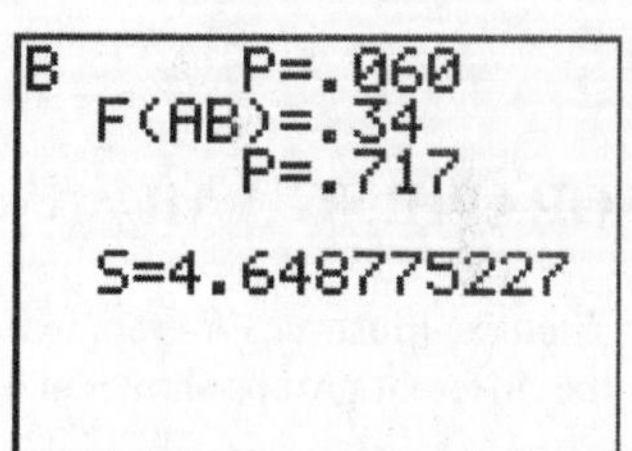

TI-89/Nspire Procedure

On these calculators we will enter the data into three lists (one for each size car), and sequentially enter the data so that all data for foreign is before that for domestic.

From the Tests menu, select option D:ANOVA2-Way, the last item on the menu. You can perform a blocked design ANOVA, or a 2-Factor with equal observations such as we have. Expand to select the appropriate number of levels of the column factor (3 for the three different car sizes), and type in the number of levels of the row factor (2 for the foreign/domestic). Press [ENTER] to continue.

The Nspire only asks for the number of different column factors at first (the number of different "treatments" within a column will be asked on the next screen).

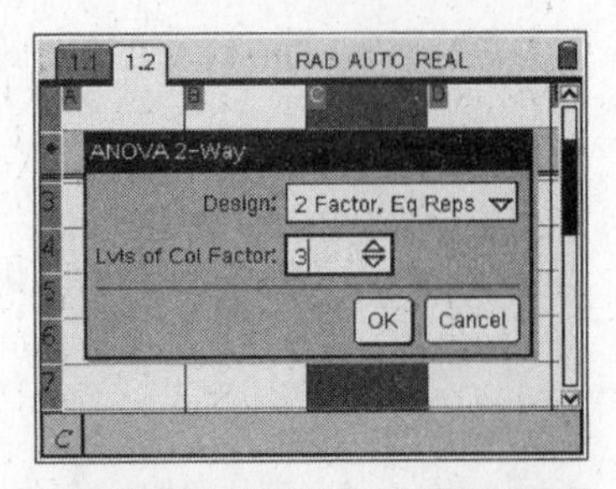

The next screen on both asks you to enter the lists which were used for the column factors (`list1` through `list3` in our example, or `a` through `c` on the Nspire). The Nspire also asks you to enter 2 for the number of different "treatments" within each column. Press [ENTER] to start the calculation and display the first portion of the results.

The first portion of the results displays the results for the column factor (the car sizes). As we saw before, the size of car comes close to being significant here with a p-value of 0.060. Press the down arrow for the rest of the output (row factor, interaction, and error).

GRAPHING A MEANS PLOT

The text illustrates a graphical way to look for differences in two-way ANOVA with a means plot. The means for the different groups shown in Table 12-3 are displayed here for convenience.

	Small Car	Medium Car	Large Car
Foreign	47.0	45.7	39.7
Domestic	43.0	38.0	36.0

To create a means plot for these data, I have entered the numbers 1 through 3 in `L1` to correspond to the three car sizes, and the means for Foreign cars in `L2`. The means for Domestic cars are in `L3`.

I have defined two connected scatter plots each of which uses `L1` as the `Xlist`. They also use different plot symbols.

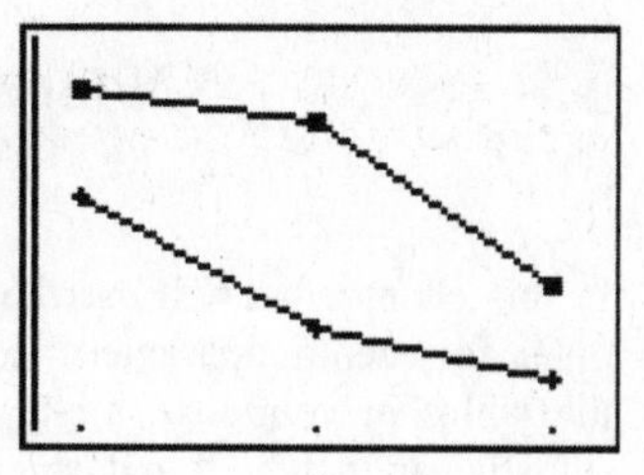

[ZOOM][9] ([F5] on an 89) will display the graph. Significant interactions are shown on these plots through dramatically nonparallel or crossing lines. Significant differences in the second (line) factor are seen by large vertical distances between the two plots. Here, we see the dramatic difference already observed for the difference in manufacture (foreign or domestic). It appears that size of car might be important as well (our *p*-value for this factor was 0.06 – close to significant). To do this on the Nspire, create a plot for one type of car, then use `Plot Properties`, `Add Y Variable` to add the second type in the graph (see the material in Chapter 1 on Ogives for more information on doing this).

Special Case: One Observation per Cell and No Interaction

Our original two-way data on car safety had three observations for each combination of size and origin. As an example, assume we have only the first entry in each cell of the table. We thus have only 6 data values (one for each size and origin combination). We will not be able to determine an effect due to interaction with only one value per cell, but we can perform two-factor ANOVA and duplicate the text's Minitab results.

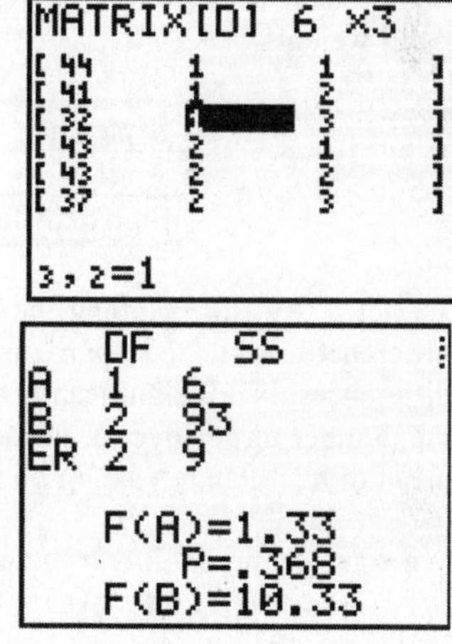

Proceed by placing the six data values into matrix D which will this time be a 6 row, 3 column matrix.

Run program **A1ANOVA** as before. The results shown are similar to the Minitab output displayed in the text. In this data set, car size matters; not whether or not the car was foreign.

TI-89 and Nspire users: proceed analogously using the 2-Way ANOVA as before, except choose `Blocked` as the type of 2-way ANOVA.

13 Nonparametric Statistics

In this chapter we will use the TI calculators' spreadsheet to perform most of the calculations. Examples will be given for each nonparametric test covered in your text. In some circumstances, when sample sizes are too large for the tables in your text, a z-test statistic which is normally distributed can be calculated. The p-value of such a statistic can then be found using the `normalcdf` function. This process will be illustrated.

In all of these tests, the procedures are analogous whether you are using an 83/84 series or an 89 or Nspire calculator. If using an 89 or Nspire, make the appropriate adjustments to list names.

SIGN TEST

Claims Involving Matched Pairs

EXAMPLE Freshmen Weight Gain? The following includes some of the data listed in Data Set 3 of Appendix B. Those weights were measured from college students in September and April of their freshman year. Use the sample data with a 0.05 significance level to test the claim that there is no difference between the September and April weights.

September	67	53	64	74	67	70	55	74	62	57
April	66	52	68	77	67	71	60	82	65	58
Sign of difference (S-A)	+	+	–	–	0	–	–	–	–	–

We let $n = 9$ (disregarding the 0 difference). If the weights were equal then the number of positive and negative differences would be approximately equal ($9/2 = 4.5$ each), but in the above table there are only 2 positive differences. Is this significantly different from what we expect? If the two weights are equivalent, the distribution of the number of positive (or negative) signs for the differences would be binomial with $n = 9$ and $p = 0.5$ yielding a mean of $np = 9*0.5 = 4.5$. (You can review the binomial distribution in Chapter 5.)

We want the probability of having 2 or fewer positive differences which is P(0) + P(1) + P(2). Press [2nd] [VARS] for the `DISTR` menu then select `binomcdf(`. Fill in the 9,0.5,2. (On the 89 or Nspire you specify the low and high ends of interest for this command.) Press [ENTER] to see the p-value of 0.0898. Since this is a two-tailed test, multiply this p-value by 2 (we get 0.1796). The p-value is much larger than 0.05, so we do not reject the hypothesis that there is *not* an increase in college students' weight during the freshman year.

Claims Involving Nominal Data

EXAMPLE Gender Selection: The Genetics and IVF Institute conducted a clinical trial of its methods for gender selection. Of 726 babies born to parents using the XSORT method to increase the probability of conceiving a girl, 668 babies were girls. Test the null hypothesis that this method of gender selection has no effect at a 0.05 significance level. Our hypotheses are H_0: $p = 0.5$ versus H_1: $p > 0.5$. In this case, the alternate hypothesis indicates that girls will be more likely to be born.

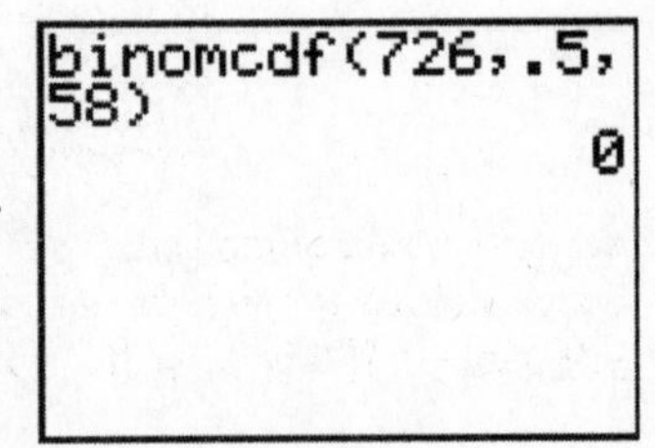

This is a binomial distribution with $n = 726$ and $p = 0.5$. The formula in the textbook is really calculating a z-score and then using the normal approximation to the binomial. With TI calculators, we can compute the p-value directly using `binomcdf`. Since there were 668 girls and 58 boys, we want the probability of 58 (or fewer) boys, which is equivalent to the probability of 668 (or more) girls. The total number of births is $n = 726$. Since the p-value essentially 0, there is enough evidence to conclude that this method does affect the gender of the babies.

Claims about the Median of a Single Population

EXAMPLE Body Temperatures: Use the sign test to test the claim that the median value of the 106 body temperatures of healthy adults (from Data Set 2 of Appendix B) is less than 98.6°F. We are thus testing

H_0: median = 98.6 versus H_1: median < 98.6

The data set has 68 subjects with temperatures below 98.6°F, 23 subjects with temperatures above 98.6°F and 15 subjects with temperatures equal to 98.6°F. (We will later present steps for how to find these values.)

Discounting the 15 temperatures equal to 98.6°F because they do not add any information to this (or any such) problem, the sample size is $n = 68 + 23 = 91$. If the median were 98.6, we would expect about half of these 91 values to be below the median and half to be above it. This is a binomial distribution with $\mu = np = 91*0.5 = 45.5$. We want the probability of having 23 or fewer values above the median (as this is what has occurred).

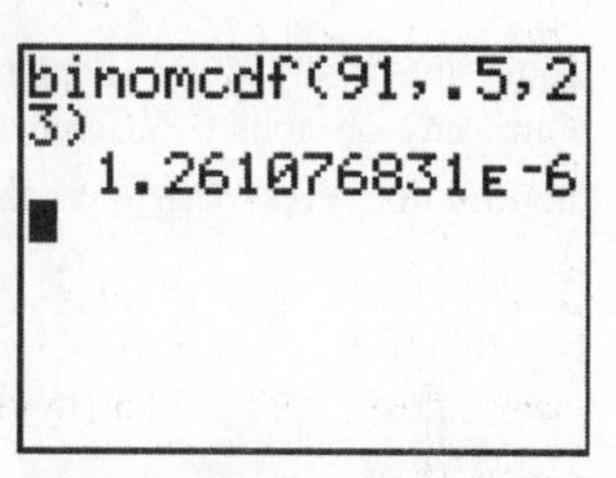

Again, we want to select `binomcdf(` from the `DISTR` menu. Fill in the 91,0.5,23. This is a symmetric distribution (under H_0), so finding 23 *above* the median is the same as finding 23 *below* the median. Press ENTER to see the *p*-value = 0.00000126 (be careful since the leading portion looks like a probability greater than 1). With such a small *p*-value there is good evidence that there are fewer temperatures above 98.6 than would be expected if it were the median. This supports the claim that the median body temperature is less than 98.6°.

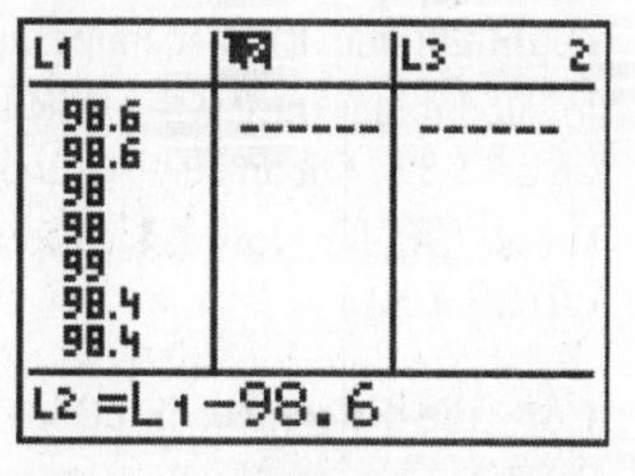

If the 106 body temperatures are saved in a list (e.g., `L1`), the following steps show one way the numbers below and above the hypothesized median could be obtained.

With your data in `L1`, highlight list name `L2`. Type `L1-98.6` as in the bottom of the screen. Press ENTER to do the calculation. We see some differences are negative, some are positive and some are 0.

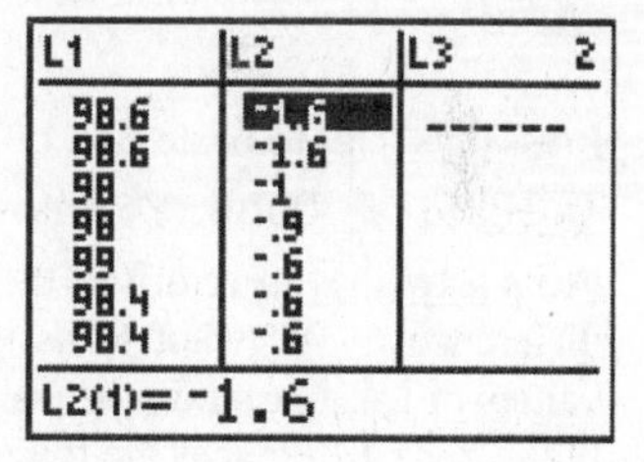

Press STAT 2 for the `SortA(`option from the `STAT` menu. (On an Nspire, find the `Sort` command as option 6 on the `Actions` menu in the Spreadsheet editor.) Press 2nd 2 to choose `L2` for sorting. Then press ENTER. The values will be sorted in ascending order when you go back to the editor.

Use the ▼ key to look at the sorted values. You will see that 68 values are below 98.6 (the differences are negative as at right). You will also find 15 values equal to 98.6 (the differences are 0) and 23 values above 98.6 (the differences are positive.) These are the numbers given earlier.

WILCOXON SIGNED RANK TEST FOR MATCHED PAIRS

EXAMPLE Freshmen Weight Gain? The following data are repeated from the first example in this chapter. We will use the Wilcoxon signed-ranks test to test the claim of no difference in weight during the Freshman year. We use significance level $\alpha = 0.05$.

September	67	53	64	74	67	70	55	74	62	57
April	66	52	68	77	67	71	60	82	65	58
Difference (S-A)	+1	+1	−4	−3	0	−1	−5	−8	−3	−1

When we look at the actual differences, we see that some are negative, some positive (there is one zero difference, which will be ignored for testing purposes); but it seems that the negative differences are *larger* than the positive differences. If this is really so, it would seem that freshmen do gain weight.

The data are from a matched pairs study, since the same students were weighed twice. From what we learned in Chapter 9, we would like to do a t-test using the differences as the data for the test. A histogram of the differences shows that these are not normally distributed. With our small sample sizes, we cannot do the *t*-test. Therefore, we must try a non-parametric test using the differences.

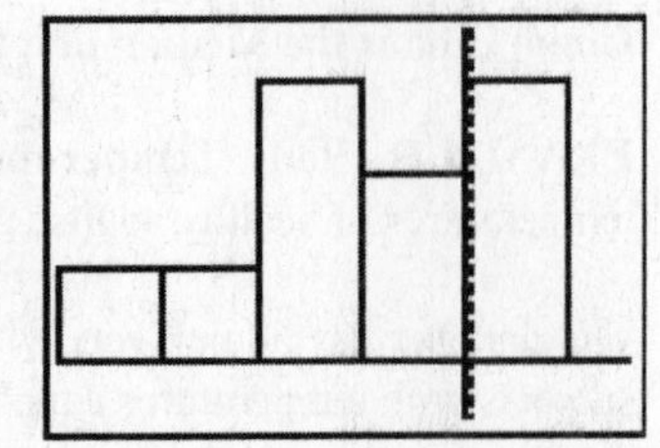

Practically speaking, for this number of data values (9, ignoring the zero difference), it might be easier to sort the differences (ignoring signs), assign ranks, and then sum the ranks for the positive and negative differences. The following steps show a method for finding the sum of the positive ranks and the sum of the absolute values of the negative ranks which is useful for larger samples.

Enter the first sample (September) in L1 and the second sample (April) in L2. Omit any obvious 0 differences. Highlight list name L3 and type L1-L2 on the bottom line. Press [ENTER] to calculate the differences.

L1	L2	L3 3
67	66	
53	52	1
64	68	-4
74	77	-3
70	71	-1
55	60	-5
74	82	-8

L3(1)=1

Highlight the L4 list name and enter abs(L3. (The absolute value function is located under the [MATH], NUM(ber) menu. It is option 1 on a TI-83/84, and option 2 on a TI-89. On an Nspire, find this on the first page of the Catalog by pressing Ⓐ.) Press [ENTER]. Now L4 will contain all positive values (the absolute value of the differences).

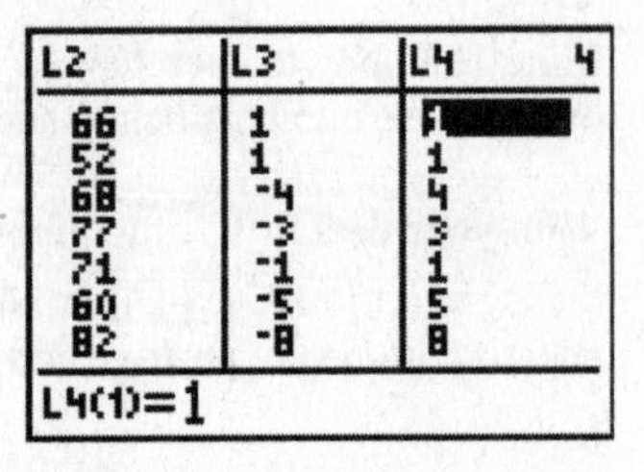

Copy the differences in L3 into L5 by highlighting L5, entering L3, and then press [ENTER].

Press [STAT] [2] to paste SortA(on the home screen. Then enter L4,L5 and press [ENTER] to see Done. You have sorted L4 and carried the contents of L5 along. Press [STAT] [1] to return to the STAT editor. L5 retains the sign of the original differences. We want to assign ranks to the differences based on the absolute values in L4, but retain the signs which are carried (for now) in L5. At this point, if any zero differences are found, delete them.

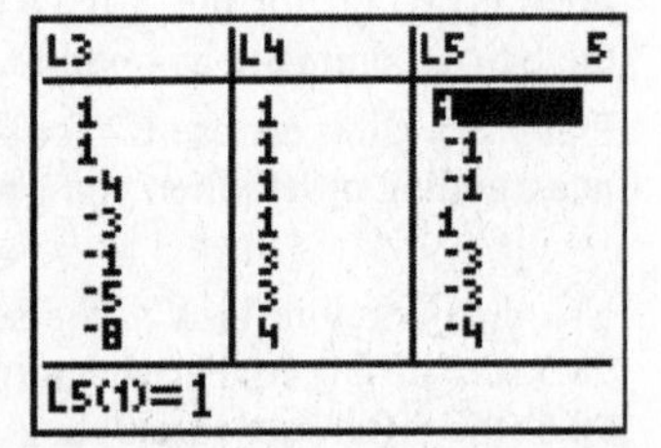

Highlight the L5 name and enter L5 [÷] L4. Press [ENTER] for a column of positive and negative ones as at right.

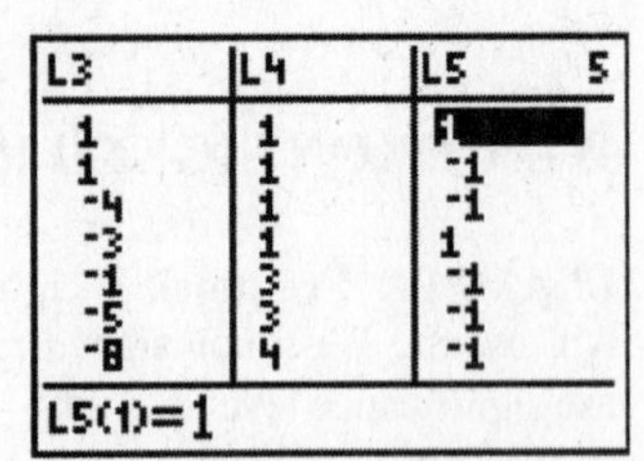

On an 80-series calculator, generate the integers from 1 to 9 (the number remaining after all zeroes are deleted) in L6. Do so by highlighting L6 and typing [2nd] [STAT] [▶] [5]. This chooses the seq(option from the LIST, Ops menu. (On a TI-89, press [F3] for the Lists menu, then [2] for Ops, then [5]). Type X,X,1,9). Press [ENTER]. At this point, you will need to scan down L4 and search for any ties. If there are any, you will need to modify L6 (the ranks) to assign an average rank for any differences which are equal. In this example, the first four differences all have

L4	L5	L6 6
1	1	2.5
1	-1	2.5
1	-1	2.5
1	1	2.5
3	-1	5.5
3	-1	5.5
4	-1	7

L6(1)=2.5

absolute value 1 (seen in L4). They all receive a rank of (1+2+3+4)/4 = 2.5. The two threes receive the average of ranks 5 and 6, and so on through the list.

To create the sequence using an Nspire, move the cursor into the formula area of a column in the spreadsheet. Press (menu), then (3) for Data, then (1) for Generate Sequence. We just want the sequence 1, 2, …, 9, so enter the command structure as in my screen at right.

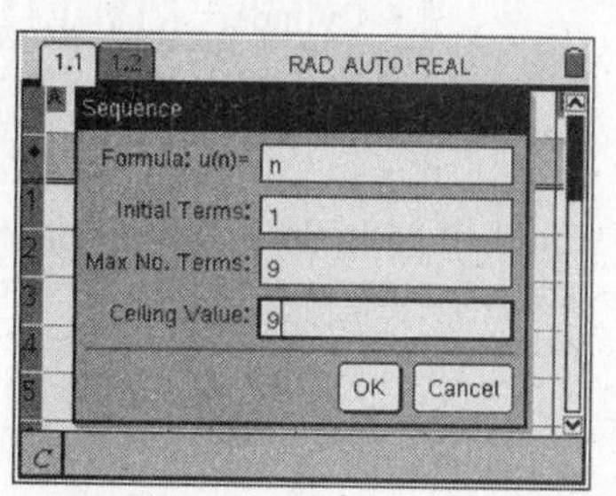

Quit and return to the Home screen. Type 9(9+1)/2 [STO▸] [ALPHA] [MATH] [ENTER] for the sum of the ranks (45) to be stored as A (Variable a on a TI-89 is [alpha][=]). As a double-check on your (possibly adjusted) ranks in L6, sum the elements in L6 with [2nd] [STAT] [▸] [▸] [5] [2nd] [6] [ENTER]. These sums of the ranks need to match. If they do not, recheck your ranks in L6.

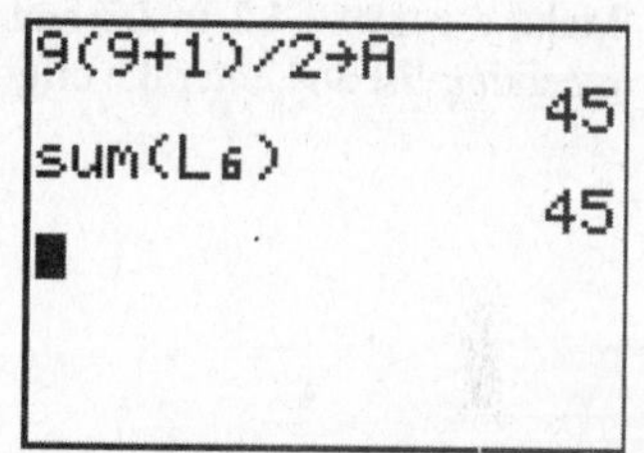

On a TI-89 from the home screen, press [2nd][5] for the MATH menu, arrow to List, press the right arrow to expand the selection and choose option 6. If you are using an Nspire, the Sum command can be found in a Calculator window on the Statistics, List Math menu.

Multiply L5*L6 and store in L6. This puts a sign on the rankings in L6.

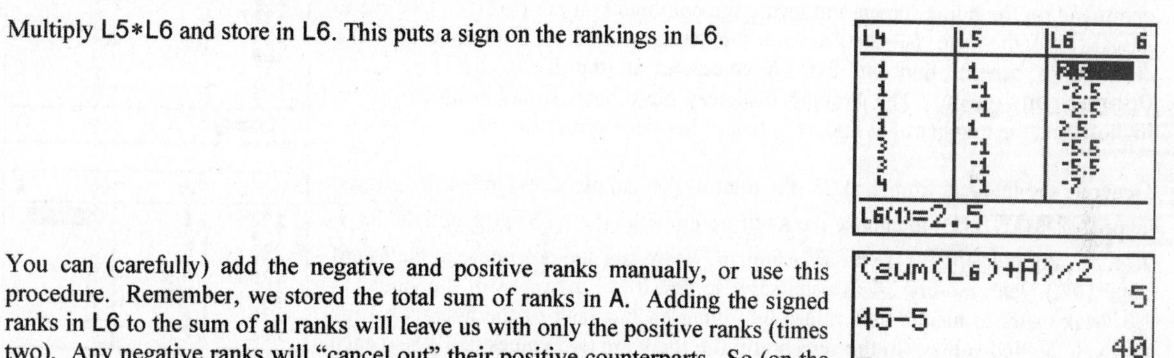

You can (carefully) add the negative and positive ranks manually, or use this procedure. Remember, we stored the total sum of ranks in A. Adding the signed ranks in L6 to the sum of all ranks will leave us with only the positive ranks (times two). Any negative ranks will "cancel out" their positive counterparts. So (on the home screen) if we enter (sum(L6)+A)/2 [ENTER], we find the sum of the positive ranks is 5. Since we know the sum of all the ranks is 45 (from above), we find that the sum of the negative ranks is 40.

```
(sum(L6)+A)/2
                5
45-5
               40
```

We use the smaller value (5) for comparison with the appropriate value from the table in the text (Table A-8). We find it is smaller than the two-tailed critical value (6), so we can reject H_0.

This test has detected a difference in median weights across the Freshman year. This test result contradicts the result from the sign test; part of the reason is that we are using more of the information in the data – not just is the weight more (or less) but adding a sense of how much. We'll add another caution to our results – these were only students at Rutgers University, and they were not a random sample, but rather volunteers; they most likely are not representative of all college freshmen.

WILCOXON RANK-SUM TEST FOR TWO INDEPENDENT SAMPLES

EXAMPLE Braking Distances of Cars: Table 13-5 lists the braking distances (in feet) of samples of 4-cylinder and 6-cylinder cars. Use a 0.05 significance level to test the claim that median braking distance is the same for both car types.

4-Cylinders	136	146	139	131	137	144	133	144	129	144	130	140	135
6-Cylinders	131	129	127	146	155	122	143	133	128	146	139	136	

The following steps give the sums of ranks for the two data sets.

Enter all 25 of the braking distances into L1. In L2 type a 1 next a 4-cylinder value and a 0 next to a 6-cylinder value. You will thus have thirteen 1's followed by twelve 0's in L2.

Make a copy of L1 in L3 and a copy of L2 in L5. (Highlight the name of the receiving list and enter the original list's name, then press [ENTER].)

Press [STAT] [2] to paste SortA(on the home screen, then type L3,L5 [ENTER] to sort the values in L3 and carry along the values in L5. On a TI-89, use this command on the home screen and locate the command on the Math, List menu ([2nd][5][3][4], then use [VAR-LINK] to enter the list names. On an Nspire, do this on a calculations screen; find the SortA command in the Statistics, List Operations menu. The braking distances have been sorted from lowest to highest, and the number of cylinders indicator has been preserved.

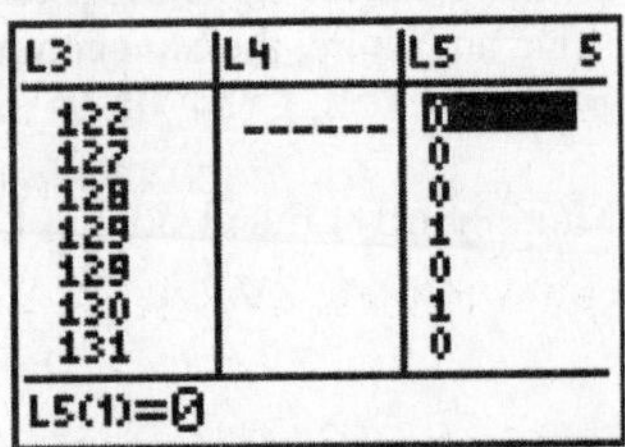

Generate the integers from 1 to 25 (the total of the sample sizes) in L4 as follows. Type [2nd] [STAT] [▶] [5] to choose the seq(option from the LIST, Ops menu. Type X,X,1,25 and [ENTER]. (If you are using an Nspire, see the procedure at the top of page 107.) Next, modify L4 as necessary, so that it has the ranks of the values in L3. Make sure to handle tied values by giving each a rank of the average of the ranks of the tied values. In the screen at right, there are two values of 129, so each is assigned a rank of 4.5; there were also two 131's that each receive rank 7.5.

Quit and return to the Home screen. Type 25(25+1)/2 [STO▶] [ALPHA] [MATH] [ENTER] for the sum of the 25 ranks which is 325, which is also stored into variable A. Find sum(L4) which should also be 325. This step verifies that you handled ties correctly.

Multiply L4 by L5 and store in L6. L6 now contains the braking distance ranks for the 4-cylinder cars. (Six-cylinder cars have a 0 in L5.)

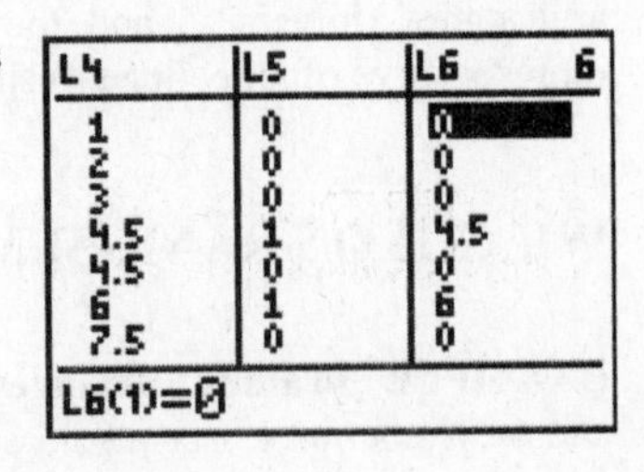

Find sum(L6 as at right. This is the sum of the 4-cylinder ranks or 180.5 = R_1 in the text. The sum of the 6-cylinder ranks is found by subtracting this value from the sum of all ranks. This value is 144.5.

```
                 325
sum(L4)
                 325
sum(L6
               180.5
A-180.5
               144.5
```

The above value for R is compared against a value from a normal distribution with mean $\mu = \frac{n_1(n_1+n_2+1)}{2} = 169$ and standard deviation $\sigma = \sqrt{\frac{n_1 n_2(n_1+n_2+1)}{12}} = \sqrt{\frac{13*12(13+12+1)}{12}} = 18.835$. The z-score of this statistic is found by the formula $z = \frac{R-\mu_R}{\sigma_R} = \frac{180.5-169}{18.385} = 0.626$. We find the area in the right tail of the standard normal distribution past $Z = 0.626$ using normalcdf(from the DISTR menu. We see the result is .2657. We double this for the two tailed test p-value of 0.5313, since the claim was that the cars had the same braking distances (we don't have prior knowledge of which direction any difference might take). We do not have evidence to reject the hypothesis that the distributions for the two samples were the same.

```
normalcdf(.626,E
99,0,1)
         .2656574091
Ans*2
         .5313148182
```

It appears that 4-and 6-cylinder cars have the same median braking distance. Note that with this small Z test statistic, we really didn't need to compute a p-value; knowing the 68-95-99.7 Rule tells us this is not an unusual Z-score.

KRUSKAL-WALLIS TEST

The Kruskal-Wallis test is an extension of the Wilcoxon Rank-Sum test. It compares the distributions from more than two independent samples, just as the ANOVA F test extends a two-sample t test.

EXAMPLE Car Crash Test Measurements: Table 13-6 (reproduced below for convenience) lists chest deceleration measurements on crash dummies for three sizes of cars. Use the Kruskal-Wallis Test to test the null hypothesis that the samples all came from populations with the same medians. Use significance level 0.05.

Small	44	43	44	54	38	43	42	45	44	50
Medium	41	49	43	41	47	42	37	43	44	34
Large	32	37	38	45	37	33	38	45	43	42

The steps below will explain how to get the sum of the ranks for the data from a specific sample.

Place all of the data values in L1. Start with the values small cars, followed those for medium cars, etc. In L2, place a 1 next to each of the values from small cars, a 2 next to each medium car value, and a 3 next to each large car value. Make a copy of L1 in L3.

```
L1    L2    L3    3
44    1     44
43    1     43
44    1     44
54    1     54
38    1     38
43    1     43
42    1     42
L3(1)=44
```

Press STAT 2 to paste SortA(onto the home screen, then type L3,L2 ENTER to sort the values in L3 carrying along L2.

```
L1    L2    L3    3
44    3     32
43    3     33
44    2     34
54    3     37
38    2     37
43    3     37
42    3     38
L3(1)=32
```

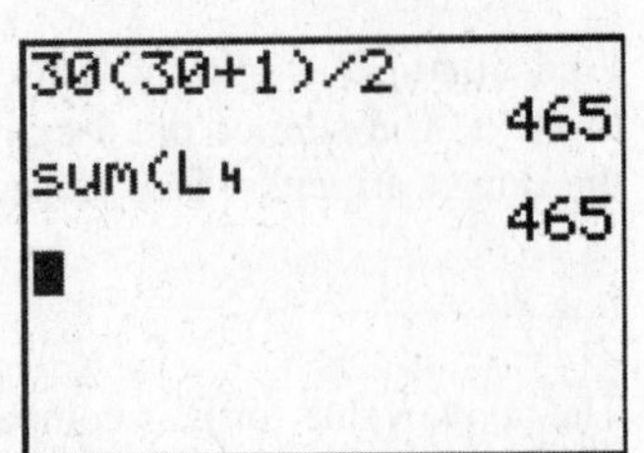

Generate the integers from 1 to 30 in L4 as follows. Type 2nd STAT ▸ 5 to choose the seq(option from the LIST, Ops menu. Type X,X,1,30 and ENTER. Next, modify L4, so that it has the ranks of the values in L3. Make sure to handle ties by assigning the average of the ranks to all tied observations.

Here, we show that the sum of ranks 1 to 30 is 465 and also confirm that the sum of the ranks in L4 is also 465. This is a double-check that tied ranks were assigned properly.

Next use SortA(L2,L3,L4 to sort the values in L2 and carry along the values in L3 and L4.

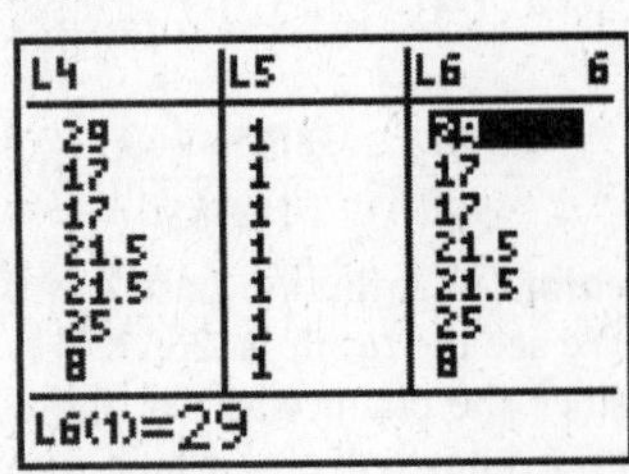

In L5, place ten 1's followed by twenty 0's. This places a 1 next to each small car rank.
Multiply L4 by L5 and store the results in L6. This stores only the small car ranks in L6. Then find sum(L6) (Find sum(on the LIST,MATH menu) to obtain R_1 = 203.5.

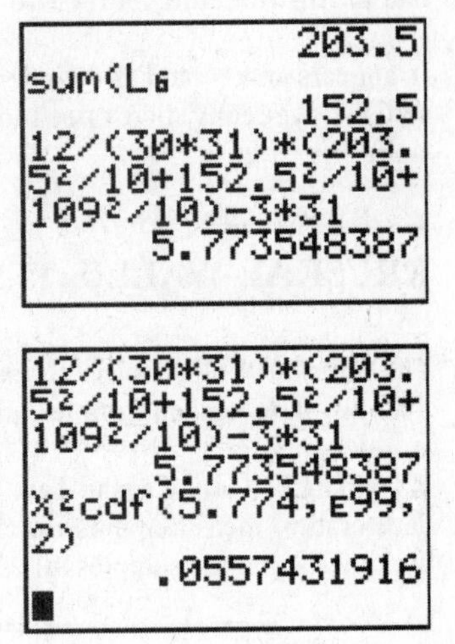

Repeat steps as above but with 1 next to the medium size ranks and 0's next to all others to get R_2 and similarly with for the large cars. Once you have determined that $R_1 = 203.5$, $R_2 = 152.5$, and $R_3 = 109$, you can calculate the test statistic H by

the formula $H = \dfrac{12}{N(N+1)}\left(\dfrac{R_1^2}{n_1}+\dfrac{R_2^2}{n_2}+\dfrac{R_3^2}{n_3}\right)-3(N+1)$

$$= \frac{12}{30(30+1)}\left(\frac{203.5^2}{10}+\frac{152.5^2}{10}+\frac{109^2}{10}\right)-3(30+1) = 5.774.$$

The distribution of H is chi-squared with $k - 1 = 3 - 1 = 2$ degrees of freedom. We find the p-value as the area above our test statistic of 5.774 using the χ^2cdf(function from the Distributions (Distr) menu. We find the p-value is 0.0557. This is more than the significance level of 0.05, so we have do not have evidence against the null hypothesis that the samples came from populations with the same medians. This sample does not find a difference in braking distance with car size.

RANK CORRELATION

Correlation and regression as studied in Chapter 10 require that the variables be numeric and that the residuals have a normal distribution around the regression line. Not all data have those properties; especially when one of the variables is a set of ranks.

EXAMPLE Are the Best Universities the Hardest to Get Into? Data in the Chapter Problem (reproduced below) lists the overall quality scores and selectivity rankings (lower values mean the school is harder to get into) for a sample of national universities (based on data from *U.S. News and World Report* magazine). Test to see if there is a correlation between the overall quality measure and selectivity rankings. Use significance level $\alpha = 0.05$.

Overall Quality	95	63	55	90	74	70	69	86
Quality Rank	8	2	1	7	5	4	3	6
Selectivity	2	6	8	1	4	3	7	5

We begin by ranking the quality scores from lowest to highest so that we now have two rank variables. (Note that we could also have ranked these in the opposite order with 1 being the most selective – that would merely change the direction of any association.) With the two sets of ranks in L1 and L2, the scatter plot (Selectivity on the y axis) shows evidence of a decreasing linear relationship between the ranks of quality and selectivity From the STAT TESTS menu, choose option `LinRegTTest`. Use the two lists as the data for the linear regression. Highlight `Calculate` in the last line, and press [ENTER].

The last line of output from the regression gives $r_S = -0.857$.

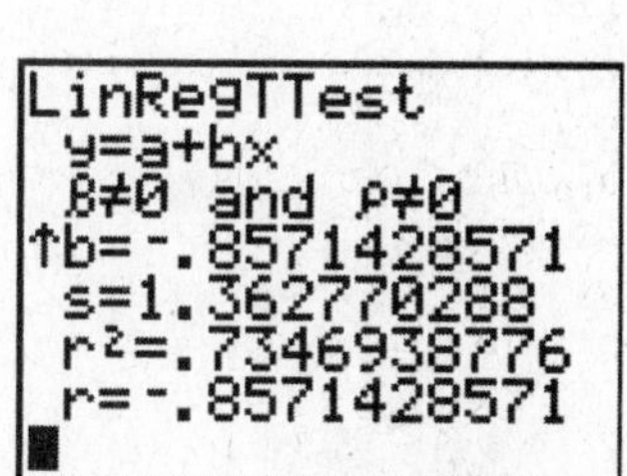

To test the hypothesis, use the critical values from Table A-9 of the text. Do not use Table A-6 (which was used for the Pearson correlation coefficient) because it requires the populations sampled to be normally distributed. Ignoring the negative sign, our statistic is larger than the .05 level critical value of 0.738 from the table.

It appears that there is a significant relationship between overall quality and selectivity of colleges; those that rank higher in quality are more selective.

Note: The t statistic and p-value given in the regression output are for the Pearson correlation coefficient and do not apply for the ranks.

RUNS TEST FOR RANDOMNESS

This is a test to decide whether (or not) data actually were randomly obtained. It is based on whether there are too few (or too many) runs of like observations in a dichotomous (only two outcomes possible) data set.

EXAMPLE Genders of Study Participants: Listed below are the genders of the first 15 subjects participating I the "Freshman 15" study with results listed in Data Set 3 in Appendix B. Use a 0.05 significance level to test for randomness in the sequence of genders.

M	M	M	M	F	M	F	F	F	F	F	M	M	F	F

There are six runs in this data: the first is MMMM, the second is F, etc. We have a total of 7 Males and 8 Females. Refer to table A-10 to find the critical value for this small data set. The critical values for sample sizes of 7 and 8 are 4 and 13. Since our number of runs (6) is between these, we can believe the sequence of genders is random.

EXAMPLE: Global Warming: Example 3 in this section of the text gives data on global mean temperatures (in °C) of the earth's surface. Each temperature represents one year, listed on order. From the text, there are 68 values above the mean of 13.998°C and 58 values below the mean. The number of runs is 32. What does this suggest about global warming?

Since $n_1 > 20$ we can use formulas given in the text for

$$\mu_G = \frac{2n_1n_2}{(n_1+n_2)} + 1 = \frac{2*68*58}{(68+58)} + 1 = 63.6032 \text{ and}$$

$$\sigma_G = \sqrt{\frac{(2n_1n_2)(2n_1n_2 - n_1 - n_2)}{(n_1+n_2)^2(n_1+n_2-1)}} = \sqrt{\frac{(2*68*58)(2*68*58-68-58)}{(68+58)^2(68+58-1)}} = 5.554.$$

Both formulas use only the values for n_1 and n_2. The test statistic z is computed as $z = (G-\mu_G)/\sigma_G = (32-63.6032)/5.554 = -5.69$.

```
2*68*58/(68+58)+
1
         63.6031746
√(2*68*58*(2*68*
58-68-58)/((68+5
8)²*(68+58-1))
         5.55449677
```

We use `normalcdf(` pasted in from the **DISTR** menu to compute the *p*-value for the test. We double the one-tailed *p*-value for the two-tailed test and obtain $1.27\times10^{-8} < 0.05$. Thus we reject the null hypothesis. There is evidence that this sequence global temperatures is not random.

```
(32-63.6032)/5.5
54
           -5.690169247
2*normalcdf(-E99
,-5.69
           1.27403E-8
```

14 Statistical Process Control

In this chapter we will use TI calculators to plot run charts and control charts for the range, $\overline{x}$, and proportions.

As with nonparametric tests, the procedures using a TI-89 and Nspire are analogous. If you have forgotten how to do these connected scatter plots, refer to Chapter 2.

RUN CHARTS

A run chart is a sequential plot of individual statistics over time (essentially a time series plot). These are used in manufacturing for several purposes – usually to monitor that a process does not stray too far from the intended mean.

EXAMPLE Global Warming: The Chapter Problem includes data on the annual global mean temperature (in °C) from 1880 with projections for 2006-2009. What does a run chart indicate about possible global warming? The data are reproduced below for convenience.

											$\overline{x}$	Range
1880s	13.88	13.88	14.00	13.96	13.59	13.77	13.75	13.55	13.77	14.04	13.819	0.49
1890s	13.78	13.44	13.60	13.61	13.68	13.68	13.73	13.85	13.79	13.76	13.692	0.41
1900s	13.95	13.95	13.70	13.64	13.58	13.75	13.85	13.60	13.70	13.69	13.741	0.37
1910s	13.79	13.74	13.67	13.72	13.98	14.06	13.80	13.54	13.67	13.91	13.788	0.52
1920s	13.85	13.95	13.91	13.84	13.89	13.85	14.04	13.95	14.00	13.78	13.906	0.26
1930s	13.97	14.03	14.04	13.89	14.05	13.92	14.01	14.12	14.15	13.98	14.016	0.26
1940s	14.14	14.11	14.10	14.06	14.11	13.99	14.01	14.12	13.97	13.91	14.052	0.23
1950s	13.83	13.98	14.03	14.12	13.91	13.91	13.82	14.08	14.10	14.05	13.983	0.30
1960s	13.98	14.10	14.05	14.03	13.65	13.75	13.93	13.98	13.91	14.00	13.938	0.45
1970s	14.04	13.90	13.95	14.18	13.94	13.98	13.79	14.16	14.07	14.13	14.014	0.39
1980s	14.27	14.40	14.10	14.34	14.16	14.13	14.19	14.35	14.42	14.28	14.264	0.32
1990s	14.49	14.44	14.16	14.18	14.31	14.47	14.36	14.40	14.71	14.44	14.396	0.55
2000s	14.41	14.56	14.70	14.64	14.60	14.77	14.64	14.66	14.68	14.70	14.636	0.36

Enter the data values in list L1. Using 1-Var Stats from the STAT, CALC menu, we find the overall mean is 14.0188. Now, generate integers from 1 to 130 and store them in L2. Highlight the list name and use the seq(option from the LIST, OPS menu and complete the command by typing X,X,1,13Ø). If you are using an Nspire, you can place your cursor in the formula box for an empty column of the spreadsheet and locate the seq(command in the ⓐ by pressing ⓢ, then arrowing down to locate the command shell. Complete the command as for the other calculators.

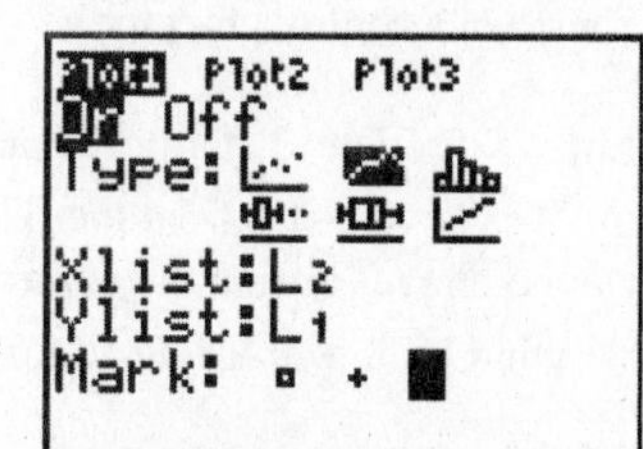

Set up Plot1 as an xy-Line plot. All other statplots should be turned off.

Press [Y=] and let Y1 = 14.0188.

Temporarily turn off the axes by pressing [2nd] [ZOOM] and choosing AxesOff. (Be sure to change this back after you are through with this chapter).

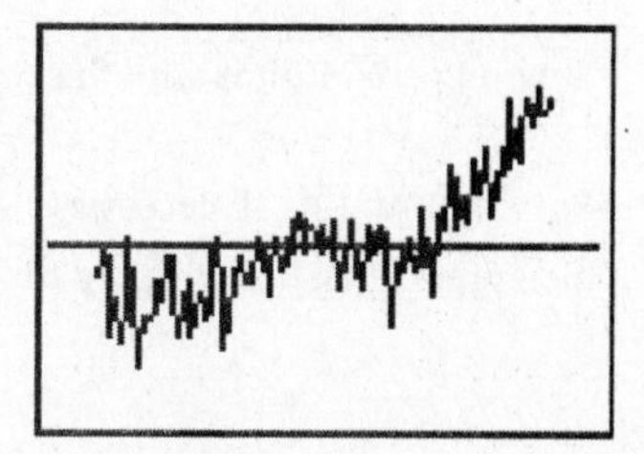

Press [ZOOM] [9] then adjust the [WINDOW] (if needed) to better fill the screen by (for this example) setting Xmin = 0 and Xmax = 130. Then press [GRAPH] to display the resized plot. To maneuver within the plot, press [TRACE] and use the right and left arrows. The reference line for the mean of the observations is clear, as is the increasing temperature as time continues.

CONTROL CHART FOR MONITORING VARIATION: THE R CHART

EXAMPLE Global Warming: Refer to the global annual mean temperatures again. If we consider each decade to be a sample, construct a control chart for R, the range of temperatures in each decade.

Put the decade numbers (integers 1 to 13) in L1 by either typing them in or using the seq(command. Enter the corresponding ranges in L2. To find the mean of the data in L2 use 1-Var Stats from the STAT, CALC menu. You will find $\overline{R} = 0.378$. From Table 14-2, we find the needed quantities to compute the control limits are $D_3 = 0.223$ and $D_4 = 1.777$ since there are ten observations in each decade. You can use $\overline{R}$ and these values to find the control limits. We find UCL = 1.777*0.378 = 0.6717 and LCL = 0.223*0.378 = 0.0843.

Set up Plot1 to be an xy-line plot as at right.

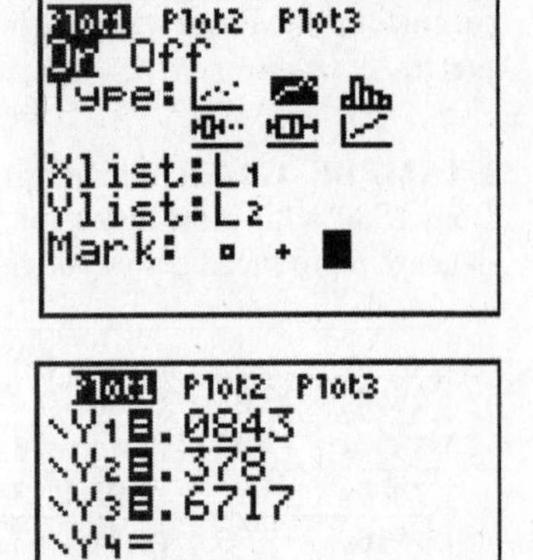

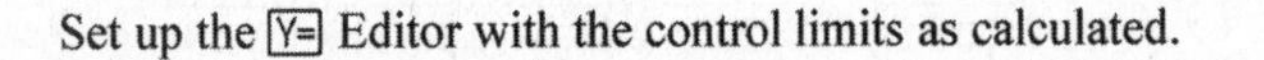
Set up the [Y=] Editor with the control limits as calculated.

Press [ZOOM] [9] then press [TRACE] for the plot which is similar to the R-chart in the text. The ranges of global temperatures by decade are well within the limits; they do not show any indications of an "out of control" situation.

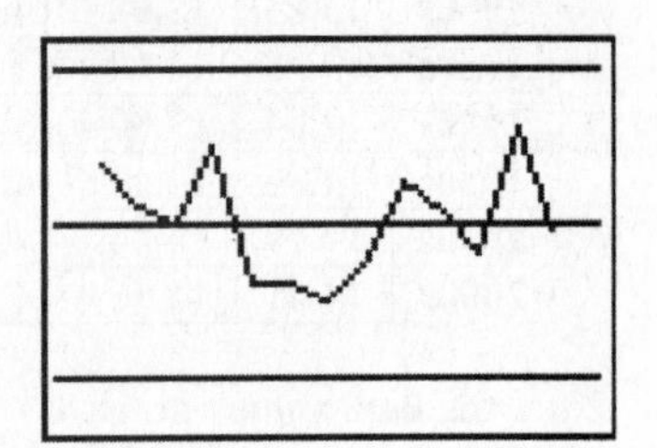

CONTROL CHART FOR MONITORING PROCESS MEAN: THE $\overline{x}$ CHART

EXAMPLE Global Warming: Refer to the annual mean temperatures data. Using the samples of size $n = 10$, construct a control chart for $\overline{x}$.

Enter 1 through 13 (for the decades) in L1 and the corresponding means in L2. To find the mean of the data in L2, use 1-Var Stats from the STAT, CALC menu. You will find $\overline{\overline{x}} = 14.019$. You can use this value and that of $\overline{R}$ (found above) to find the control limits as in the text, using the constant from Table 14-2 (here, $A_2 = 0.308$). You will find UCL = $\overline{\overline{x}} + A_2\overline{R} = 14.019 + 0.308*0.378 = 14.135$ and LCL = $\overline{\overline{x}} - A_2\overline{R} = 14.019 - .308*.378 = 13.903$.

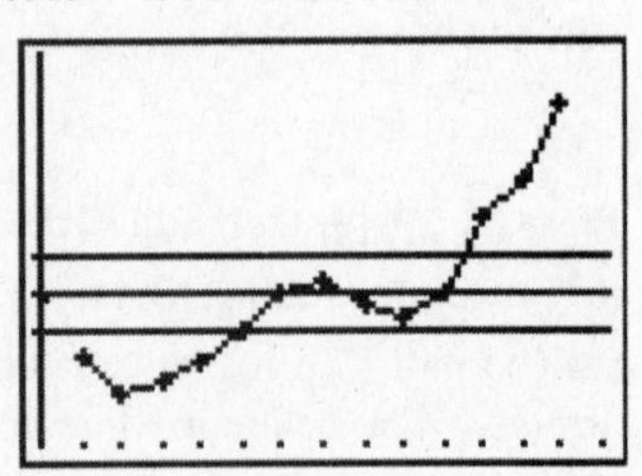

Set up Plot1 to be an xy-line plot, as has been done previously.

Set up the [Y=] Editor with the control limits as was done for the range plot above.

Press [ZOOM] [9]. If necessary, change the [WINDOW] so that both limits are visible, then press [GRAPH] to display the plot which is similar to the $\overline{x}$ -chart in the text.

CONTROL CHART FOR ATTRIBUTES: THE P-CHART

EXAMPLE Defective Heart Defibrillators: The Guidant Corporation manufactures implantable heart defibrillators. Families of people who have died using these devices are suing the company. According to *USA Today*, "Guidant did not alert doctors when it knew 150 of every 100,000 Prizm 2DR defibrillators might malfunction each year." Because lives could be lost, it is important to monitor the manufacturing process of such devices. Listed below are the numbers of defective defibrillators in successive batches of 10,000. Construct a control chart for the proportion p of defective defibrillators and determine whether the process is within statistical control.

Defects	15	12	14	14	11	16	17	11	16	7	8	5	7	6	6	8	5	7	8	7

Enter the batch number in `L1` (integers from 1 to 20; remember you can use the `seq(` command to do this easily) and the corresponding number of defectives in `L2`. On the Home screen, type `L2/10000` and store in `L2`. This will replace the numbers of defectives with the proportion of defectives.

Set up `Plot1` as in the preceding examples.

Set up the [Y=] editor with the control limits as has been done previously. These limits are calculated using $\overline{p}$ (use `1-Var Stats` to find the average of the proportions, which is 0.001) as $\overline{p} \pm 3\sqrt{\frac{\overline{p}\overline{q}}{n}}$ which give 0.0766 and -0.0216 (use 0 for any values less than 0). Remember that n is the number per batch, not the number of batches.

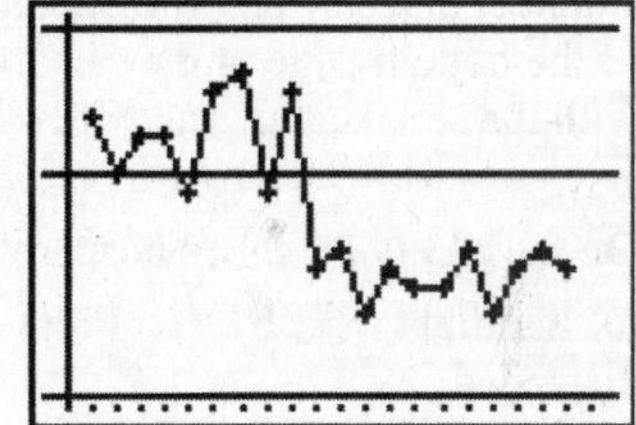

Press [ZOOM] [9]. Change `Ymin` and `Ymax` if necessary so that both the LCL and UCL are visible. Use [TRACE] as necessary to fully examine the graph. We notice a dramatic decrease in the proportion of defectives; while the process is in control, the company might want to reevaluate their limits based on a recent history of fewer defectives (which is good).

Appendix

This appendix contains information on transferring data and programs from the CD-ROM that came with your main text. Also, we give instruction on how to install a data set from the data Apps and how to make and install your own group.

The CD-ROM contains the data sets from Appendix B of *Elementary Statistics* (11th Edition) by Mario F. Triola. For the TI-83+ and TI-84 calculators, data is given in an App (or application) called TRIOLAXE.APP. For the TI-83/89, data is given as individual lists in ASCII format as text files (with extensions of .txt). The CD-ROM also contains two programs A1ANOVA.83p (used in Chapter12) and A2MULREG.83p (used in Chapter 10).

Your instructor will probably load the data (and programs, if needed) onto your TI calculator, or you can transfer them from the CD-ROM with your computer if you have TI-Connect or TI-GRAPH LINK software and cable available from Texas Instruments (the cable and software are included with the TI-84, as well as the TI-89 Titanium edition). See the guidebook that comes with the calculator for information on the TI-GRAPH LINK.

With the exception of .fig files created with Cabri II Plus, TI-Nspire Computer Link Software allows only TI-Nspire handheld documents to be copied to a handheld. If you inadvertently try to copy other types of files, or a folder containing other types of files, TI-Nspire™ Computer Link Software notifies you and cancels the copy operation. (This means that these means of transferring files won't work with an Nspire.)

Loading Data or Programs from One TI-83/84 to Another

Connect one calculator to another with the cable that came with the calculator. The link port is located at the center of the bottom edge of the calculator on the 83 series, on the top for the 84 series (which can communicate with 83's with the serial cable, and with other 84s with the USB cable as well).

On the receiving calculator, press [2nd] [X,T,Θ,n] to choose the LINK menu. Press [▸] to highlight RECEIVE. Press [ENTER] or [1] and see the message Waiting displayed.

```
SEND RECEIVE
1:All+...
2:All-...
3:Prgm...
4:List...
5:Lists to TI82...
6:GDB...
7↓Pic...
```

On the sending calculator, press [2nd] [X,T,Θ,n]. Press the down arrow to scroll through the list of options. When the category of what you wish to send is highlighted, press [ENTER]. In this example we are sending choice C:Apps, but we could send a program or a list of data with options 3 and 4. Once we press [ENTER] to select option C:Apps, we use [▾] to locate the TRIOLAXE App and press [ENTER] to select it.

```
SEND RECEIVE
8↑Matrix...
9:Real...
0:Complex...
A:Y-Vars...
B:String...
C:Apps...
D↓AppVars...
```

Press [▸] to highlight TRANSMIT. Press [ENTER] or [1] to transmit whatever you have chosen on the previous step.

Note: If you are transmitting an App the receiving calculator will signal "garbage collecting" then "receiving", and then "validating" before indicating "Done", so be patient. If the name of whatever is being sent is already in use on the receiving calculator that calculator will show a screen like the one at right. You can then choose to overwrite the old with the new or to rename so you can keep both.

```
DuplicateName
1:Rename
2:Overwrite
3:Overwrite All
4:Omit
5:Quit
  FPULS    LIST
```

USING THE DATA APPS (TI-83+/84 ONLY)

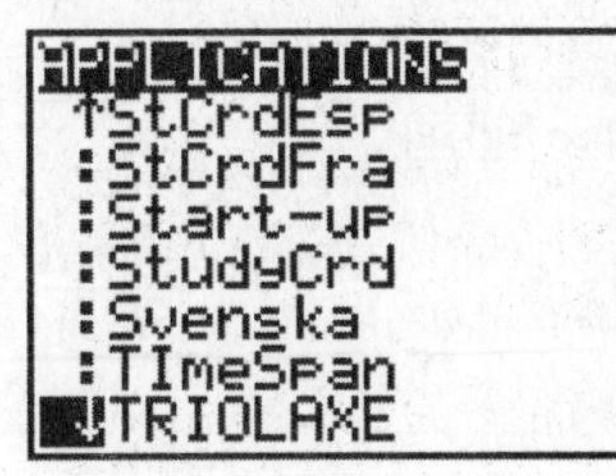

Press the [APPS] key for a list of all applications which have been installed on the calculator. (We assume that TRIOLAXE App has been installed in this example. The list of other Apps available may vary.) Use [▼] to locate the TRIOLAXE App.

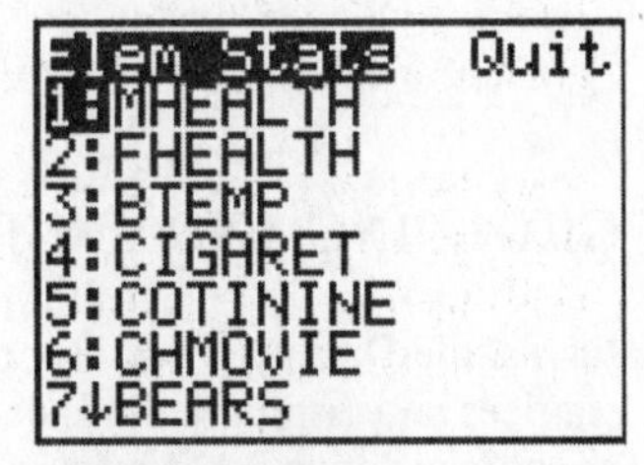

Then press [ENTER] for a title screen that soon changes to this screen which contains a list of the data sets in Appendix B of your text.

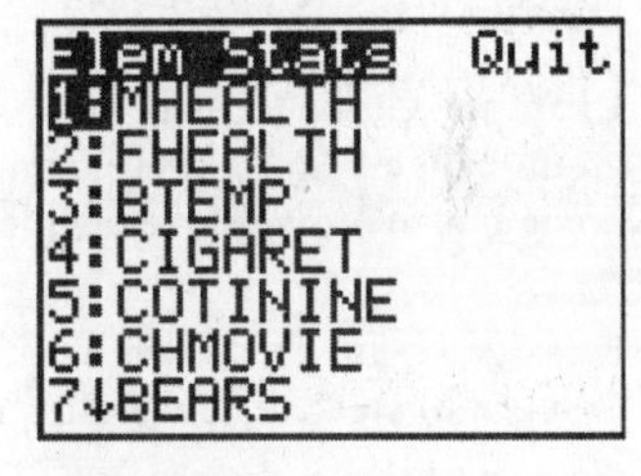

Press [2] to choose FHEALTH and see a list of the data lists from the data set FHEALTH.

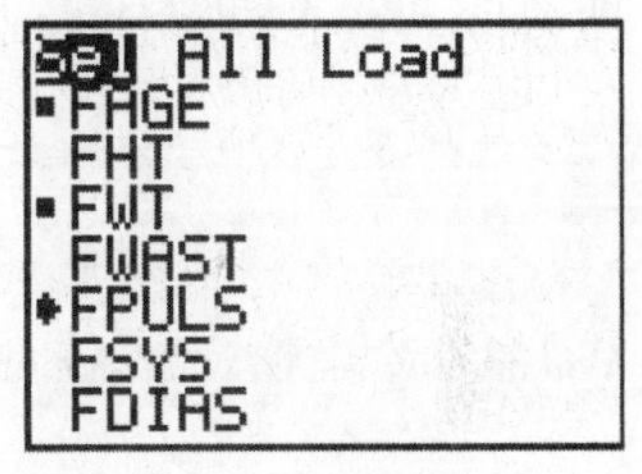

Cursor down and press [ENTER] next to each list you wish to load. A small black square will designate a chosen list. For our example we choose FAGE, FWT and FPULS. If you wish to load all lists in a set then press [▶] to highlight "All".

Press [▶] [▶] to highlight Load. At this screen, press [1] to choose SetUpEditor.

Note: The option 2:Load loads the lists from archive memory to random access memory. The data would not be loaded to your Statistics Editor but would be available on the Lists menu by pressing [2nd] [STAT].

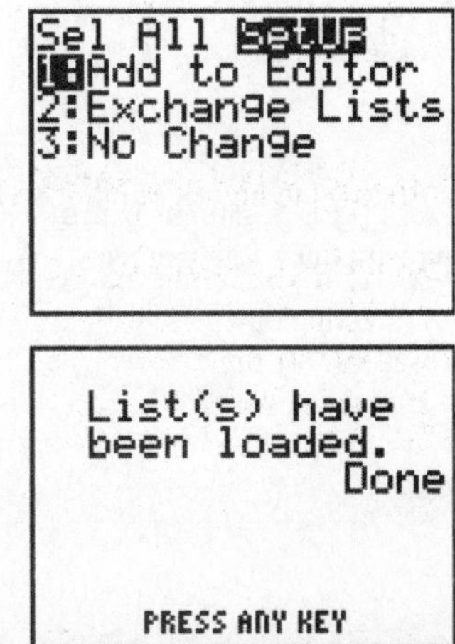

At this screen, press [2] to choose the "Exchange Lists" option.

Note: Option 1 will add the lists to the lists already in the Stats Editor. Option 3 will not change the editor and place the lists in the LIST menu as described in the Note above.

After making your selection, you will be returned to the previous screen. Now, select option 2:Load.

This screen informs you that the List(s) have been loaded. Pressing any key returns you to the initial data set selection screen. Pressing [2nd] [MODE] lets you QUIT and return to the Home screen (or press the right arrow and then [ENTER]).

To see the results, press [STAT] [1] to check out the Stat Editor. Here we see the lists we loaded are in fact there and ready for us to use.

FAGE	FWT	FPULS 1
17	114.8	76
32	149.3	72
25	107.8	88
55	160.1	60
27	127.1	72
29	123.1	68
25	111.7	80

FAGE(1) =17

GROUPING AND UNGROUPING (TI-83+/84 ONLY)

Just as the Data App has lists of data from the text grouped together, you may want to group lists of data and/or matrices and/or programs. The advantage to doing this is that groups are saved in Archive memory and do not take up room in active RAM memory until you want to use them.

Example Group the data sets FAGE, FWT, and FPULS just moved in from the App in the preceding example into a group called FEM. (These data are already part of the group FHEALTH in the TRIOLAXE App, so we are just doing this to provide an example.)

Press [2nd] [+] to choose the MEM menu. Then choose option 8:Group. Press [ENTER] to select to create a new group. You will be prompted to name your group. Type in the name FEM and press [ENTER].

You are now asked what should be in the group. Press [4] to choose List.

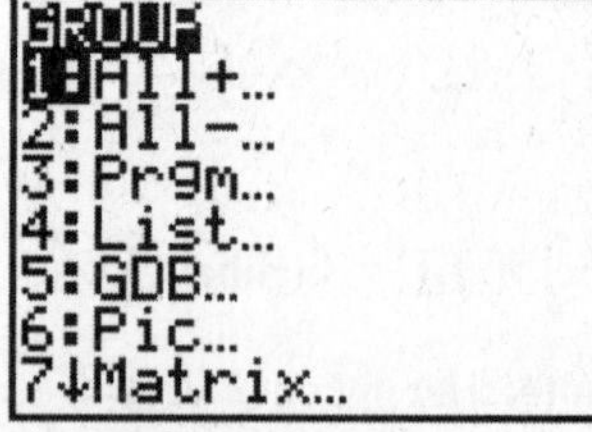

Use the cursor to move down the list of data lists and press [ENTER] next to each list which you wish to include in the group FEM. Again, chosen lists will be designated by little black squares in the left margin.

Press [▸] to highlight Done. Then press [ENTER] and you will see this completion screen after a few moments.

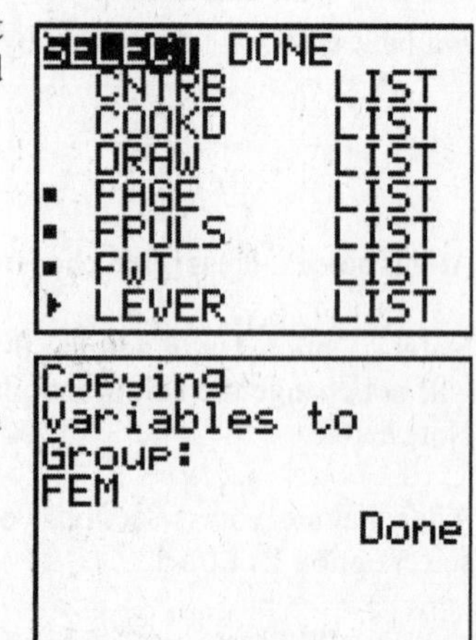

Now if you delete the three lists from RAM they will still be in archive memory. To return them to RAM, return to Group/Ungroup screen (option 8 from the MEM menu) and highlight Ungroup and press [ENTER]. You would then select group FEM to ungroup. Even though the list will then reside in RAM, it will remain in archive memory unless it is deleted as a Group using option 2:Mem Mgmt/Del from the MEM menu.

USING .TXT FILES (ALL 8x SERIES MODELS)

The CD-ROM that is included with the text has data files in text (.txt) format which can be loaded into the calculators without retyping the data, if you have the proper computer cable and TI-Connect software. While using this procedure is probably not time effective for small data sets, it can save both time (and the aggravation of typing mistakes) with larger ones. The method is simple for anyone who is familiar with basic cut-and-paste editing on a computer.

Insert the CD, and select the Datasets folder. Then, select the textfiles folder. There are icons and short names for each data set (list). Select the data set you wish and double click it to open it (the computer will use Notepad as the default software). Drag the mouse cursor to highlight the list contents. Press Control-C to copy the data onto the computer's scratchpad.

Start the TI-Connect software, and select the TI DataEditor application. Click on the blank page icon. You will see a list with a zero (Ø) as the first entry. Click the cursor in the first cell, and press Control-V (Paste) to place the list contents into the editor. You will need to give your list a "name" and TI-calculator properties. Click File, Properties (Control-R). Select the appropriate device type and give the list a name. If you are using a TI-89 series calculator, you will also be asked the folder you want the data placed in. Click the OK button. Now, click the Send File icon on the menu bar (it looks like a TI calculator with an arrow pointing toward it); the transfer should start automatically. If the list name is already in use (L1 through L6, for example), you will be asked whether you wish to abort the transfer, overwrite the current contents, or give a new name to the list.